# Decoding Schopenhauer's Metaphysics

The key to understanding how it solves
the hard problem of consciousness and
the paradoxes of quantum mechanics

# Decoding Schopenhauer's Metaphysics

The key to understanding how it solves
the hard problem of consciousness and
the paradoxes of quantum mechanics

Bernardo Kastrup

IFF
BOOKS

Winchester, UK
Washington, USA

JOHN HUNT PUBLISHING

First published by iff Books, 2020
iff Books is an imprint of John Hunt Publishing Ltd., No. 3 East Street, Alresford,
Hampshire SO24 9EE, UK
office@jhpbooks.com
www.johnhuntpublishing.com
www.iff-books.com

For distributor details and how to order please visit the 'Ordering' section on our website.

ISBN: 978 1 78904 426 3
978 1 78904 427 0 (ebook)
Library of Congress Control Number: 2019941535

A CIP catalogue record for this book is available from the British Library.

Design: Stuart Davies

UK: Printed and bound by CPI Group (UK) Ltd, Croydon, CR0 4YY
US: Printed and bound by Thomson-Shore, 7300 West Joy Road, Dexter, MI 48130

We operate a distinctive and ethical publishing philosophy in
all areas of our business, from our global network of authors to
production and worldwide distribution.

# Contents

# Other books by Bernardo Kastrup

*Rationalist Spirituality: An exploration of the meaning of life and existence informed by logic and science*

*Dreamed up Reality: Diving into mind to uncover the astonishing hidden tale of nature*

*Meaning in Absurdity: What bizarre phenomena can tell us about the nature of reality*

*Why Materialism Is Baloney: How true skeptics know there is no death and fathom answers to life, the universe, and everything*

*Brief Peeks Beyond: Critical essays on metaphysics, neuroscience, free will, skepticism and culture*

*More Than Allegory: On religious myth, truth and belief*

*The Idea of the World: A multi-disciplinary argument for the mental nature of reality*

The perfected masterpiece of a truly great mind will always have a profound and vigorous effect on the whole human race, so much so that it is impossible to calculate to what distant centuries and countries its enlightening influence may reach.
Arthur Schopenhauer, in *The World as Will and Representation* (1818)

We philosophers ... are no thinking frogs, no objectifying and registering devices with frozen innards—we must constantly give birth to our thoughts out of our pain and maternally endow them with all that we have of blood, heart, fire, pleasure, passion, agony, conscience, fate, and disaster. Life—to us, that means constantly transforming all that we are into light and flame, and also all that wounds us ... Only great pain is the liberator of the spirit.
Friedrich Nietzsche, in *The Gay Science* (1882)

# Chapter 1

# Introduction

*Before we can discern the new, we must know the old. The adage that everything has already happened, and that there is nothing new under the sun (and the moon), is only conditionally correct. It is true that everything has always been there, but in another way, in another light, with a different value attached to it, in another realization or manifestation.*
Jean Gebser, in *The Ever-Present Origin* (1966)

Born in Danzig—present-day Gdańsk—to German-Dutch parents in 1788, Arthur Schopenhauer gained recognition as a philosopher only in the last decade of his life, in the mid-19th century. His main work, *The World as Will and Representation*, came to light precisely 200 years before I started writing the present book.

Today, Schopenhauer is best known for his psychology, ethics, aesthetics and prose style. When it comes to metaphysics, however, his philosophy has been considered "so obviously flawed that some people have doubted whether he really means it" (Janaway 2002: 40). This is tragic, for I believe Schopenhauer's most valuable legacy is precisely his metaphysical views: they anticipate salient recent developments in analytic philosophy, circumvent the insoluble problems of mainstream physicalism and constitutive panpsychism, and provide an avenue for making sense of the ontological dilemmas of quantum mechanics. As I shall soon argue, there is certainly nothing "obviously flawed" about his views; much to the contrary. Had the coherence and cogency of Schopenhauer's metaphysics been recognized earlier, much of the underlying philosophical malaise that plagues our culture today—with its insidious effects on our science, cultural ethos and way of life—could have been avoided.

With the present book, I hope to contribute to changing this regrettable state of affairs. In the pages that follow, I offer a conceptual framework—a decoding *key*—for interpreting Schopenhauer's metaphysical arguments in a way that renders them mutually consistent and compelling. With this key in mind, it is my hope that even those who have earlier dismissed Schopenhauer's metaphysics will be able to return to it with fresh eyes and at last unlock its sense.

I admittedly interpolate Schopenhauer's assertions—i.e. I fill in the gaps in his argument—in a manner that some may consider too, well, creative. Let me acknowledge upfront that I may, *in some sense*, be guilty of this. In my defense, however, I offer the following contention: if one (re-)reads Schopenhauer's words *under the light of the interpretation elaborated upon here*, one will find it difficult to imagine that Schopenhauer could have meant anything substantially different from what I posit. So let my interpretation be judged not by the wording of isolated passages of Schopenhauer's writings, but by how well it brings Schopenhauer's *overall* metaphysical argument together in a coherent, unifying and clarifying way.

I only truly discovered Schopenhauer's metaphysics after having fleshed out my own views on the nature of reality; a decade-long effort—totaling seven books—completed with *The Idea of the World*. I thus brought to bear on my read of Schopenhauer an extensive preexisting background of related ideas and insights.

Two inferences could then reasonably be made from this confession: first—and on a positive note—that my own work equipped and primed me for discerning the intended meaning of Schopenhauer's contentions, despite his relatively loose and seemingly contradictory use of words. After all, I had just spent years wrestling with the same problems he wrestled with, working out similar solutions, and could thus not only understand but also *recognize* Schopenhauer's contentions. Second—and this time on a negative note—it could also be argued that my prior metaphysical work imparts a structural bias in my efforts to interpret Schopenhauer: I am primed to read into his words a reflection of my own views.

Both inferences probably have some merit. Let me highlight, however, that throughout the writing of this book I have been aware of this inherent potential for bias and made deliberate efforts to avoid it. As much a reflection of persisting partiality as this very statement could still represent, I believe my analysis and conclusions are fairly objective. Readers should be able to assess whether this is or isn't the case based on how well I substantiate my argument in the pages that follow.

Another confession: Schopenhauer initially attracted me because of his ethics, his way of dealing with the sufferings of life, not his metaphysics. I began my exploration of his thought with Christopher Janaway's little book, *Schopenhauer: A Very Short Introduction*. In it, Janaway introduces Schopenhauer's ethics by first summarizing its metaphysical basis, the foundation upon which Schopenhauer builds the edifice of his broad philosophical system. In the many quotes of Schopenhauer's works included in the book, I believed to discern—to my surprise—clear similarities with the metaphysics laid out in my own work. Naturally, I felt his points were compelling.

Yet, Janaway peppered his book with criticisms of Schopenhauer's metaphysics. What he seemed to be making—or failing to make—of Schopenhauer's words was quite different from what I thought to discern in them. Janaway saw problems and contradictions where I thought to see clarity, elegance and consistency. But since Janaway is the professed expert and I was just perusing quotes out of context, I initially suspected I was reading too much into them.

The only way to clarify the issue was to sink my teeth into

3

Schopenhauer's *magnum opus*: the two-volume, 1,200-page-long third edition of *The World as Will and Representation*, in the same translation that Janaway himself used. Although Schopenhauer wrote a few other books discussing more specific topics, *The World as Will and Representation* stands as his only work of systematic philosophy (Young 2017), comprising the main articulation of his metaphysics.

In the ensuing months, I devoured the lengthy two-volume set, reading and re-reading it. I recognized in it numerous echoes and prefigurations of ideas I had labored for a decade to bring into focus. The kinship between my own work and what I was now reading was remarkable, down to details and particulars. Here was a famous 19th century thinker who had already figured out and communicated, in a clear and cogent manner, much of the metaphysics I had been working on. What better ally could I have found? And yet, bewilderingly to me, Schopenhauer's "metaphysics has had few followers" (Janaway 2002: 40). Its utter failure to impact on our culture for the past 200 years is self-evident to even the most casual observer.

The present volume is thus a product of both dismay and delight: dismay at how misunderstood Schopenhauer's metaphysics seems to be, even at the hands of presumed experts; and delight at the discovery that my own metaphysical views have such a clear and solid historical precedent.

My goal with this book is thus two-fold: on the one hand, I aim to rehabilitate and promote Schopenhauer's metaphysics by offering an interpretation of it that resolves its apparent contradictions and unlocks the meaning and coherence of its constituent ideas. On the other hand—and on a more self-serving note—I hope to show that my own metaphysical position, as articulated in my earlier works, isn't peculiar or merely fashionable, but part instead of an established, robust and evolving chain of thought in Western philosophy.

As an important bonus, by showing that Schopenhauer's metaphysics can be coherently interpreted in a way that reveals how much it has in common with my own, I also indirectly situate my work in the context of earlier Western thinkers, such as Spinoza, Berkeley, Kant and Hegel, as well as Eastern philosophical traditions. After all, Schopenhauer himself explicitly situated his metaphysics in that broader context.

It is critical that those who hope to truly understand Schopenhauer do *not* expect from him the kind of rigorous, consequent, consistent use of terms that is today characteristic of analytic philosophers. Needless to say, Schopenhauer preceded analytic philosophy by a century. His intended denotations of key terms are *context-dependent*. He may, for instance, use the term 'consciousness' in the sense of explicit or meta-cognitive awareness in one context, and then in the sense of mere experience in another. Analogously, he may use the verb 'to know' in the sense of true cognition in one context, and then in the sense of mere experiential acquaintance in another. And so on.

Indeed, to understand Schopenhauer's metaphysics one must read him *charitably*, always looking for the particular one, amongst the various possible denotations of a term, which fits most coherently into his overall scheme. The interpretational flexibility this requires is familiar to every non-philosopher in everyday conversation: despite often-loose use of words by one's interlocutor, one knows what is meant because of the context. Indeed, what makes Schopenhauer so delightful to read is precisely that he writes in a colloquial manner—as if he were trying to verbally explain something to the reader in person—so we must reciprocate and interpret him with equally colloquial flexibility. This is perfectly feasible because Schopenhauer is delightfully verbose: he repeatedly recapitulates and summarizes—using different words and constructs—what he has already said.

The argument in the present book thus relies on a context-dependent interpretation of Schopenhauer's use of terms. Based on it, I shall argue that the key to resolving the seeming internal contradictions of Schopenhauer's metaphysics lies in understanding the difference between phenomenal consciousness and what is today called 'meta-consciousness' — or 'conscious meta-cognition' — in psychology. I shall elaborate on this difference, show that Schopenhauer explicitly leverages it throughout his argument, and then explicate how it reconciles his seemingly conflicting metaphysical claims.

I shall also attempt to bring out the overall sense and coherence of Schopenhauer's metaphysics by placing his key contentions in an overarching conceptual framework, built upon the notion of psychological dissociation. I shall substantiate this framework with present-day psychiatric literature on Dissociative Identity Disorder, a condition in which individuals manifest multiple disjoint centers of consciousness.

On a more general note, the present volume marks an attempt by me to return to my original writing style: brief, parsimonious, to-the-point expositions. In other words, I've tried to keep this book short, no space being wasted on related but ancillary ideas — let alone divagations and digressions — so it can be read comfortably in a weekend.

My objective in doing so is *not* to oversimplify things or acquiesce to the demands of a culture of intellectual laziness — readers will soon notice that I may be guilty of many sins, but not this particular one — but, instead, to maintain focus and improve clarity. I prefer to be effective in conveying one key message than to be ineffective in addressing a variety of supporting or related ideas. The price of this frugality, however, is that this book requires attention from its readers: sometimes a crucial point is made in a single short paragraph, whose importance is disproportional to its length and can easily be overlooked in a casual read.

Still in the spirit of focus and parsimony, I shall restrict myself as much as possible to only two key sources: the Payne translation of Schopenhauer's *The World as Will and Representation* (1969), volumes 1 and 2—which I shall henceforth cite simply as 'W1' and 'W2,' respectively—and Christopher Janaway's *Schopenhauer: A Very Short Introduction* (2002). The latter I shall use as the source of present-day criticisms and objections to Schopenhauer's metaphysics, which I'll then attempt to refute. As for the former, because I aim to show that much of what I claim in the present book can be traced back to Schopenhauer's own words, I shall quote frequently from it. The many other entries in the bibliography are relatively ancillary, cited not to open up new fronts of argument, but simply to provide a more robust substantiation for my interpretation of *The World as Will and Representation*.

The focus on the two key sources mentioned above prevents me, of course, from further addressing the vast amount of secondary literature available today on Schopenhauer. For this reason, some may consider the present book less than scholarly. If so, so be it. Reviewing a multitude of secondary analyses doesn't seem—to me, at least—indispensable for accurately discerning what the primary work itself has to say: one assumes that its author is the whole and ultimate authority when it comes to his own message.

*The World as Will and Representation* is Schopenhauer's key articulation of his ontology, while Janaway's *Schopenhauer: A Very Short Introduction* is probably the only text many students of philosophy today will ever read about Schopenhauer's thought. The former is the primary source regarding Schopenhauer's metaphysics, whereas the latter is arguably the most representative example of how that metaphysics is, in my view, misunderstood today. Contrasting the two is thus significant in and of itself, notwithstanding the remaining literature.

It is my hope that the present volume contributes original and interesting views on Schopenhauer's thought, despite—or perhaps precisely *because* of—its focus and parsimony.

# Chapter 2

# Brief overview of Schopenhauer's metaphysics

*[O]ur vital energy comes from a Will which is wild, unprincipled, amoral ... a universe which is not necessarily structured and limited by a rational, benign plan, one where we cannot touch bottom, but which is nevertheless the locus of our dark genesis. ... Something which comes from the depths has its own numinosity ... The primitive has power, on which we need to draw, or before which we stand in awe, even as we may have to limit it, resist it.*
Charles Taylor, explaining the ethos of Schopenhauer's philosophy in *A Secular Age* (2007)

Schopenhauer's metaphysics is characterized by a partition of the world into two categories, which he calls 'will' and 'representation,' respectively. The latter is the outer appearance of the world: the way it presents itself to our observation. The former, on the other hand, is the world's inner essence: what it is in itself, independently of our observation.

This partition may *superficially* resemble a form dual-aspect theory (Atmanspacher 2014); indeed, at the time of this writing, Wikipedia listed Schopenhauer's metaphysics as an instance thereof. According to dual-aspect theory, mentality and physicality are two different aspects or views of the same underlying, fundamental 'stuff' of nature, which in turn is neither mental nor physical. Whether we apprehend this fundamental 'stuff' through its physical or mental aspect is a question of perspective or point of view. Those who consider Schopenhauer's metaphysics an instance of dual-aspect theory equate will with mentality and representation with physicality.

There is, however, no mention or hint in Schopenhauer's

argument of anything that could constitute an ontological ground underlying both will and representation; no mention or hint of anything that will and representation could be mere aspects *of*. The only unifying ontological claim Schopenhauer makes is that everything is intrinsically *will*, representation being merely how the will *presents* itself to observation. As he puts it, the will "is the being-in-itself of *every thing* in the world, and is *the sole kernel of every phenomenon*" (W1: 118, emphasis added), whereas representation is merely the "will become visible" (W1: 107) or "translated into perception" (W1: 100). For Schopenhauer, representations without underlying will would be "like an empty dream, or a ghostly vision not worth our consideration" (W1: 99). There is nothing more fundamental than the will, the "inner nature" (W1: 97) of everything, for, as Schopenhauer repeatedly affirms, "The will itself has no ground" (W1: 107). It is thus at least difficult to see how dual-aspect thinking, as it is formally defined in philosophy, could be attributed to Schopenhauer.

Schopenhauer is, in fact, an *idealist* with regard to the physical world—i.e. the world of material objects interacting with one another in spacetime, according to causal laws. For him, this physical world exists only insofar as it consists of mental images—representations—in the consciousness of the observing individual subject. It has no existence beyond this individual subject. Schopenhauer writes that

> things and their whole mode and manner of existence are inseparably associated with our consciousness of them. ... the assumption that things exist as such, even outside and independently of our consciousness, is really absurd. (W2: 9)

A 'thing' for Schopenhauer is a physical *object* with a certain form, occupying a position in spacetime and obeying causal laws. Unambiguously, he claims that the demand for the existence of the object outside the representation of the subject ... has no meaning at all, and is a contradiction ... therefore, the perceived world in space and time ... is absolutely what it appears to be (W1: 14)

That the physical world is what it appears to be means that it is made of *qualities* such as color, tone, flavor, odor, etc.—i.e. *it is constituted by experiential states*[1] *of the observing individual subject.* And that's all there is to it. There is no consciousness-independent physical world, comprising separate objects with definite form, physical properties and position in spacetime, which somehow correspond isomorphically to our perceptual experience. According to David Chalmers' classification scheme of variants of idealism (2018), Schopenhauer's metaphysics can thus be considered a form of 'subjective idealism' in regard to the physical world.

But Schopenhauer doesn't stop here. He posits that 'behind' the representations—i.e. 'behind' the *physical* world—there lies the *world-in-itself*, which is "completely and fundamentally different" (W1: 99) from what appears on the screen of our perception.[2] This world-in-itself is what remains of the world when it is not being observed—i.e. when it is not being represented in the consciousness of an individual subject. The "forms and laws" ordinarily discernable through perception "must be wholly foreign" to the world as it is beyond representation (*Ibid.*). In other words, the world-in-itself is *not* physical; in it there is no space, time or causality, which are themselves merely modes of perception (W1: 119-120).

The question that then arises is: What is the essential nature—the categorical basis—of the world-in-itself? Schopenhauer describes it repeatedly as *volitional states*—such as an "irresistible impulse," "determination," or "keen desire" (W1: 118)—which implies that the world-in-itself is mental. And although representations are also mental, the experiential states that constitute the world-in-itself are completely different from the qualities of representation. After all, what it feels like to desire or fear is completely different from what it feels like to perceive.

Surprisingly to me, there has been controversy about what Schopenhauer means by the word 'will.' Janaway, for instance, believes that

> we must enlarge its sense at least far enough to avoid the barbarity of thinking that every process in the world has a mind, a consciousness, or a purpose behind it. (2002: 37)

Nonetheless, I shall argue in this book that the world-in-itself, according to Schopenhauer, is indeed *mental*—i.e. constituted by experiential states, even though states very different from perceptual ones. If I am correct, Schopenhauer's position in regard to the world-in-itself fits into Chalmers' 'objective idealism' (2018).

In summary, Schopenhauer's world-in-itself is essentially mental, which implies objective idealism in regard to it. But the experiential states constituting the world beyond ourselves need not have any qualitative similarity whatsoever with the colors, tones, flavors, etcetera that we experience when observing such world. In other words, what it feels like to *be* the universe surrounding us is rather different from what it feels like to *perceive* such universe. The experiential states underlying the world we inhabit are separate from, and at least ordinarily inaccessible to, us as individual observers; all we can access is their *representations*. The latter—which constitute what we call the 'physical world'—exist only insofar as *we* experience them as individual subjects. This implies subjective idealism in regard to the physical world.

I shall later clarify all this in more detail. For now, the

important point is that Schopenhauer's metaphysics isn't a form of dual-aspect theory, but *idealist* through and through: it entails both subjective idealism—the physical world of objects in spacetime existing only as images in an individual subject's consciousness—*and* objective idealism—the world-in-itself being constituted by volitional experiential states.

## Chapter 3

# Our portal to the world

*Here I am in the presence of images, in the vaguest sense of the word, images perceived when my senses are opened to them, unperceived when they are closed. ... The afferent nerves are images, the brain is an image, the disturbance travelling through the sensory nerves and propagated in the brain is an image too. ... To make of the brain the condition on which the whole image depends is in truth a contradiction in terms, since the brain is by hypothesis a part of this image.*

Henri Bergson, in *Matter and Memory* (1896)

Schopenhauer divides representations into two contrasting categories: intuitive and abstract. Abstract representations correspond to conceptual reasoning, thus originating in the individual subject's own mind. Intuitive representations, in turn, originate in the perception of an external world. Haldane and Kemp even translate the original German *"intuitiven Vorstellung"* (Schopenhauer 1859: §3) as "idea of perception" (Schopenhauer, Haldane & Kemp 1909: §3).

Schopenhauer explicitly associates intuitive representation with perception. For instance, already in his opening definition, he writes that intuitive representation "embraces the whole visible world, or the whole of experience" (W1: 6). Since vision is a category of perception, the reference to the visible world implies perception. Moreover, Schopenhauer routinely uses the word 'experience' in the restrictive sense of conscious *perception*. He "determines experience as the law of causality" (W1: 7), which, for Schopenhauer, is our means to "logically organize our *field of sensations*" (Wicks 2017, emphasis added).

The passage that perhaps most succinctly establishes the

intended meaning of these various notions with respect to one another is this:

> The *concepts* [i.e. abstract representations] form a peculiar class, *existing only in the mind of man*, and differing entirely from the *representations of perception* [i.e. intuitive representations] ... It would therefore be absurd to demand that [the concepts] should be demonstrated in *experience*, in so far as we understand by this the real *external world* that is simply representation of perception (W1: 39, emphasis added)

Although intuitive representations or perceptions are the individual subject's portal to the external world, what Schopenhauer means by 'perception' is *more* than sense impressions alone—i.e. more than the data gathered through the mediation of the five senses. To see it, consider this more complete quote of Schopenhauer's definition: intuitive representation "embraces the whole visible world, or the whole of experience, *together with the conditions of its possibility*" (W1: 6, emphasis added). These conditions of the visible world's possibility are space and time, without which it could not be perceived.

Following Immanuel Kant, Schopenhauer claims that space and time can be "directly perceived" even "by themselves and separated from their content" (W1: 7). This implies that a person in an ideal sensory deprivation chamber still 'perceives'—i.e. somehow becomes experientially acquainted with—spacetime, although she sees, hears, smells, tastes and touches nothing. In other words, Schopenhauer posits that we have internal access to spatio-temporal extension independently of the five senses. This "*a priori* perception" (*Ibid.*) of spacetime is even a prerequisite for perception proper: intuitive representations must couch sense impressions in endogenous spatio-temporal extension.

For Schopenhauer, spacetime is thus *an internal cognitive scaffolding inherent to the individual subject,* whereas sense impressions are projected onto—and thereby populate—this scaffolding as *mental images.*

In addition, perception for Schopenhauer also entails the immediate, non-deliberate recognition of cause-and-effect relationships in sense data. We don't just experience raw sensory input—otherwise we would see the world as a chaotic flow of disconnected, senseless pixels without recognizable entities or patterns of behavior—but project a structure onto it that organizes the corresponding sense data into "comprehensible and interrelated objects" (Wicks 2017), acting on one another according to causal laws. This—together with the underlying cognitive spatio-temporal scaffolding—is a function of what Schopenhauer calls the 'understanding' or the 'intellect,' terms he uses somewhat more restrictively than we do today.

So perception for Schopenhauer entails *cognitive processes* operating below the ordinary reach of introspection or explicit awareness. Although this is entirely consistent with how perception is regarded in modern psychology (cf. e.g. Bernstein 2010), one could argue that it blurs the boundary between intuitive and abstract representations, insofar as we equate the former with perception. After all, *both* now entail cognitive processes.

However, there is in fact no such blurring. The cognitive processes entailed by perception are autonomous and cannot—at least ordinarily—be accessed through introspection. Schopenhauer uses the qualifier 'immediate' to characterize them as such. Conceptual reasoning, on the other hand, is deliberate and introspectively accessible, therein lying the defining difference between the two. I shall elaborate on this in more detail later.

For the sake of clarity, since the qualifier 'intuitive' nowadays has denotations very different from—even contrary to—what Schopenhauer intended, I shall henceforth refer to intuitive representations as 'perceptual representations.'

## Chapter 4

# The world as it is in itself

*[For historically early man] there stands behind the phenomena, and on the other side of them from me, a represented which is of the same nature as me. Whether it is called 'mana', or by the names of gods and demons, or God the Father, or the spirit world, it is of the same nature as the perceiving self, inasmuch as it is not mechanical or accidental, but psychic and voluntary.*
Owen Barfield, in *Saving the Appearances* (1957)

For Schopenhauer, the world-in-itself is *will*, all perceptual representations being 'objectifications' of this will—i.e. renditions of the will in the form of mental images in the consciousness of an individual subject. As such, the world *presents*[3] itself to us as images projected onto our internal cognitive spatio-temporal scaffolding of perception. Philosopher Itay Shani calls these mental images the "revealed order" (2015) and I the "extrinsic appearance" (Kastrup 2018a) of the world. But the world as it is in itself—i.e. its "concealed order" (Shani 2015) or "intrinsic view" (Kastrup 2018a)—is, according to Schopenhauer, something qualitatively quite different.

One can only become acquainted with the intrinsic view— the concealed order—of an aspect of the world by *being* this aspect. For without *being* it, one can only know it through how it presents itself in perception. There is, therefore, precisely *one* aspect of the world whose intrinsic view we can access: *ourselves.* What it is like to be ourselves is, for Schopenhauer, our sole hint to what the world-in-itself is like. In his wonderfully aphoristic words, "we must learn to understand nature from ourselves, not ourselves from nature" (W2: 196).

The rationale behind this central idea of Schopenhauer's requires some elaboration. According to him, one can only talk of a plurality of individual entities or events in the context of spacetime extension: two entities or events are only separate insofar as they occupy different positions in space or time. Two stones existing in the present moment can only be said to be separate if one is here and the other is there. Two events unfolding in the same place can only be said to be separate if one occurs after the other. If you and I occupied the exact same volume of space at exactly the same time, we would overlap with one another and effectively be one.

Schopenhauer calls the carving out of the world into individual entities and events, as enabled by spacetime extension, the *"principium individuationis"* (W1: 112), or 'principle of individuation.' Without spacetime extension, all entities and events would overlap and become indistinguishable from one another; the whole world would become one indivisible, dimensionless whole.

Now, as we've seen, according to Schopenhauer the spatio-temporal scaffolding is a *cognitive feature of the intellect*; it exists only in the consciousness of the individual observing subject. Consequently, the partitioning of the world into a plurality of separate entities and events can also only exist in the intellect, in the form of representation. It is *us*, in perceiving the world, who break up its image into distinct pieces. Plurality is imposed by *us*, as a mode of our cognition, which motivates Schopenhauer to refer to it as a mere "illusion" (W2: 321). The world as it is in itself, beyond representation, is outside spacetime and can, therefore, only be a unitary whole.

Since we and the world are then ultimately one, there must be a sense in which what it is like to be us—after we set all representations in our intellect aside—is akin to what it is like to be the world as a whole. By virtue of *being* ourselves we can then make inferences about the inner essence of the world.

But this key epistemic claim of Schopenhauer's—namely, that we can know something essential about the world at large merely by introspecting—doesn't rely exclusively on the idea of extra-spatiotemporal unity. Even if we grant that the world 'behind' representation is a collection of separate entities and events, an empirical argument can still be made: whatever we find out about our inner essence through introspection, our body, as it is represented on the screen of perception, is made of matter. Schopenhauer generalizes this observation by stating, "matter is that whereby ... the inner essence of things ... becomes perceptible or *visible*" (W2: 307, original emphasis). And since the world at large is—just like our body—*also made of matter*, we have reason to infer that the world at large is, in essence, also whatever it is we are, in essence. Assuming otherwise would entail postulating an arguably arbitrary discontinuity in nature. After all, the world at large is made of the same kinds of atoms and force fields our body is made of.

It is thus the empirical observation that all things perceived are made of matter—*including our body*—that still allows us to extrapolate our knowledge of our own inner essence towards the world at large, even when assuming the latter to be fundamentally constituted by separate—or at least separa*ble*—entities and events.

Whether we come to it through the notion of extra-spatiotemporal unity or through inferring that "Matter is the visibility of the will" (W2: 308), the key epistemic insight underlying Schopenhauer's metaphysics is that we can "understand nature from ourselves, not ourselves from nature" (W2: 196). This key insight is what allowed Schopenhauer to reach far beyond the limits of Kant's philosophy.

Through introspection, what Schopenhauer realized about his own inner essence—and, therefore, about the inner essence of the world as a whole—is something he considered appropriate to call the 'will.' This is a clear reference to volitional feelings. Moreover, as we've seen above, the Schopenhauerian thing-in-itself is only knowable by gauging what it is like to *be* it. And since Thomas Nagel's seminal 1974 paper—titled *What is it like to be a bat?*—philosophers have understood that what it is like to be something is the very *definition* of phenomenal consciousness (Block 1995, Chalmers 2003). For both these reasons, the world-in-itself, according to Schopenhauer, must be *experiential* in nature.

But because this conclusion leads to *seeming* contradictions and implausibilities in Schopenhauer's metaphysics—which I shall elucidate later—there are persisting doubts about it in the literature. I confess to feeling nonetheless dismayed at these doubts, for—to paraphrase Michael Tanner in an entirely analogous discussion (2001)—if by 'will' Schopenhauer meant something other than will, why didn't he then call it what he meant?

It's not like Schopenhauer is obscure in this regard: the will is "what is known *immediately* to everyone" (W1: 100, emphasis added) and "Consciousness alone is *immediately* given" (W2: 5, emphasis added).[4] So the will can only be (volitional) consciousness. Indeed, only experiential states can be known immediately. Nothing else can, for everything else is only accessible through the mediation of representation.

Schopenhauer directly associates the will with consciousness:

what as representation of perception I call my body, I call my will in so far as *I am conscious of it* in an entirely different way ... the body occurs *in consciousness* in quite another way, *toto genere* different, that is denoted by the word *will* (W1: 102-103, emphasis added)[5]

Clearly, the will consists of experiential states.

Even in his extensive dismissal of solipsism—which he

calls "theoretical egoism"—Schopenhauer uses the terms "mere phantoms" and "phenomena of the will" in reference to philosophical zombies and conscious organisms, respectively (W1: 104), thereby again equating the will with consciousness.

And as if all this weren't enough, at one point Schopenhauer refers to the will as "the inner, simple consciousness"[6] that constitutes "the one being" of nature (W2: 321). How could he be clearer?

A particular type of experiential state is more primarily associated with the will: after defining 'feeling' as "something *present in consciousness* [but which] is not a concept" (W1: 51, emphasis added), Schopenhauer claims that

> virtue and holiness [i.e. forms of conduct] result not from reflection, but from the *inner depth of the will* ... Conduct, as we say, happens in accordance with *feelings* (W1: 58, emphasis added)

So at least some of the experiential states we call 'feelings' are the same thing as—or at least very intimately related to—the inner depth of the will. Indeed, Schopenhauer repeatedly identifies feelings with the will. For instance, he says that "the inner nature of the world [i.e. the will] ... expresses itself intelligibly to everyone in the concrete, that is, *as feeling*" (W1: 271, emphasis added).

Despite all this, Janaway still claims that

> When I am conscious of my own willing in action, what I know is a phenomenal manifestation of the will, *not the thing in itself*. (2002: 39, emphasis added)

This conclusion cannot be true at least in some important sense, for after listing pleasure and pain as examples of feelings (W1: 51), Schopenhauer proceeds to distinguish them from any kind of representation:

> we are quite wrong in calling pain and pleasure representations, for they are not these at all, but *immediate affections of the will* (W1: 101, emphasis added)

So if some feelings aren't representations, either Schopenhauer is postulating a third category in his metaphysics—which would contradict his defining claim that the world is nothing but will and representation[7]—or we have to understand some feelings as the thing in itself in some sense. In other words, these immediate affections of the will must *be* the will in action. Contrary to Janaway's conclusion, there must thus be at least some important sense in which we, when consciously experiencing some of our feelings, know the will itself. Indeed, according to Schopenhauer, "the will ... shows itself as terror, fear, hope, joy, desire, envy, grief, zeal, anger, or courage" (W2: 212), endogenous feelings we are all directly acquainted with. I shall return to this later.

The difficulty here is the same apparent contradiction that permeates Schopenhauer's metaphysics: if the thing in itself lies outside spacetime,[8] it cannot be known insofar as knowledge must extend across our cognitive spatio-temporal scaffolding. Yet, Schopenhauer is definite when he claims that not only is the will known to us, it is "infinitely better known and more intimate than anything else" (W2: 318), for we can access it directly through our intrinsic view—our first-person experience—of ourselves.[9] To make sense of this apparent paradox is what I shall now begin to attempt.

## Chapter 5

# Phenomenal consciousness and meta-consciousness

*The struggle between the specifically human and the universally natural constitutes the history of man's conscious development.*
Erich Neumann, in *The Origins and History of Consciousness* (1949)

To gain introspective access to an experience—i.e. to be able to report the experience to oneself—it is not enough to merely *have* the experience; one must also consciously know *that* one has it—i.e. one must become *explicitly aware* of it by placing one's *attention* on it. This conscious knowledge *of* the experience—which comes *in addition* to the experience itself—is what Jonathan Schooler calls a 're-representation':

> Periodically *attention* is directed towards explicitly assessing the contents of experience. The resulting *meta-consciousness* involves an explicit *re-representation* of consciousness in which one interprets, describes, or otherwise characterizes the state of one's mind. (Schooler 2002: 339-340, emphasis added)

Although re-representation is necessary for *introspection*, research has shown that it is largely absent, for instance, in ordinary dreams (Windt & Metzinger 2007). This demonstrates compellingly that mental activity does *not* need to be re-represented in order to be *experienced*—after all, dreams *are* experienced—but only to be introspectively accessed. *During* ordinary dreams one simply experiences, without consciously knowing *that* one experiences; introspective access to dream contents becomes possible only upon awakening, through memory. Arguably, all animals experience their lives without necessarily re-representing their experiences.

Re-representations are the product of a *self-reflective configuration* of consciousness, whereby the latter turns in upon itself so to objectify its own contents (Kastrup 2014). In humans, this often occurs through the use of 'semiotic mediation' (Valsiner 1998), which is our ability to re-represent our experiences by *naming* them explicitly or implicitly. Alex Gillespie gives an example:

> In order to obtain dinner one must first name … one's hunger … This naming, which is a moment of self-reflection, is the first step in beginning to construct, semiotically, a path of action that will lead to dinner. (2007: 678)

So there is a close relationship between what Schooler calls 'meta-consciousness' and linguistic modes of thinking. That we can name and then *report* our inner experiences to others and ourselves—i.e. that we can tell others and ourselves *that* we have the experiences—is a hallmark of meta-consciousness. The corresponding re-representations are meta-cognitive reflections of lower-level contents of consciousness.

In philosophy of mind, a similar analysis has been offered by Ned Block (1995) through his notions of 'phenomenal consciousness' (or 'P-consciousness') and 'access consciousness' (or 'A-consciousness'). For Block, P-consciousness entails *experiential states*—that is, states in which there is something it is like to be. A-consciousness, in turn, entails what he calls "representational contents," whereby the subject's mind points to, or denotes, other ones of its own states. P-conscious states are, of course, not necessarily A-conscious, for they don't need to denote other mental states. But, conversely, A-conscious states also aren't necessarily P-conscious: in principle, a mental state could denote another without being itself experiential.

In Block's terminology, what Schooler calls 'meta-consciousness' entails an extension of P-consciousness into A-consciousness, so to produce states that are *both* P-conscious *and* A-conscious (I shall call these PA-conscious states): first, the subject has purely P-conscious states; then, by *re-representing* these original P-conscious states, the subject acquires PA-conscious states that enable introspective access, reasoning and speech.

It is the reflections of the original P-conscious states in PA-conscious states that give rise to our sense of self as know*ing* *subjects* (experiencing the reflections) separate from their know*n* *objects* (the reflected lower-level experiences). After all, we don't identify with what we perceive or feel, but with that which knows *that* it perceives and feels: we are not the table we see or the anxiety we feel, but that which knows *that* it sees a table and feels anxiety. Without the reflections, there is a sense in which the distinction between *feeling* the anxiety and *being* the anxiety would vanish; there would be no observer *of* the anxiety, but only the anxiety itself, as felt.

By re-representing our own mental activity, we thus create an experiential know*er*-know*n* pair, which is characteristic of self-reflection. The process can even lead to a recursive hierarchy of subjectivity: in a higher level of self-reflection, what was the know*er* in a lower level can itself be reflected and become the know*n* to a meta-knower.

It is crucial to keep in mind that meta-consciousness is *different* from — even though it presupposes — consciousness. A being is conscious if there is something it is like to *be* the being — i.e. if the being has experiential states. But these experiential states do not necessarily need to be re-represented. Indeed, nothing prevents experiences from occurring outside the field of self-reflection. Gregory Nixon, for instance, calls these "unconscious experiences" (2010: 216), which in my view is an oxymoron but illustrates the subtlety of the point.

Moreover, the emergence of so-called 'no-report paradigms' in contemporary neuroscience attests to the abundant presence of waking experiences that are unreportable to self or others because they fall outside the field of self-reflection (Tsuchiya *et al.* 2015, Vandenbroucke *et al.* 2014). David Eagleman provides several examples, such as the driver who moves his foot halfway to the brake pedal before becoming explicitly aware of danger ahead (2011).

Unfortunately, many conflate consciousness proper with *meta*-consciousness. Ap Dijksterhuis and Loran Nordgren, for instance, "define conscious thought as object-relevant or task-relevant cognitive or affective thought processes that occur while the object or task is *the focus of one's conscious attention*" (2006: 96, emphasis added). They insist, "it is very important to realize that *attention is the key* to distinguish [*sic*] between unconscious thought and conscious thought. *Conscious thought is thought with attention*" (*Ibid.*, emphasis added). In appealing to *attention*, as opposed to raw experience or qualia, they implicitly restrict consciousness to self-reflection.

Even more strikingly, Axel Cleeremans explicitly *defines* consciousness as self-reflection. He overtly conflates experience with meta-consciousness and reportability:

Awareness, on the other hand, always seems to minimally entail the ability of knowing *that* one knows. This ability, after all, forms the basis for the verbal reports we take to be the most direct indication of awareness. And when we observe the absence of such ability to report on the knowledge involved in our decisions, we rightfully conclude that the decision was based on *unconscious* knowledge. Thus, it is when an agent exhibits knowledge of the fact that he is sensitive to some state of affairs that we take this agent to be a *conscious* agent. This *second-order* knowledge, I argue, critically depends on learned systems of *meta representations*, and forms the basis for *conscious experience*. (Cleeremans 2011: 3, emphasis added)

This isn't a recent problem. Already in his important 1995 paper, Block lists—and is very critical of—similar conflations (pp. 236-239). Going even further back, by perusing the original texts of the founders of depth psychology one quickly realizes that, when they spoke of 'unconsciousness,' the founders often meant lack of *meta*-consciousness—not of experience proper. This is abundantly evident, for instance, in an essay written by Carl Jung in the early 1930s, titled "The Stages of Life" (2001: 97-116), in which he talks about children slowly attaining 'consciousness' as they grow up. Naturally, Jung couldn't have meant by this that newborns and toddlers lack experiential states.

As I shall elaborate upon shortly, having clarity about the distinction between consciousness and *meta*-consciousness—a distinction I shall maintain that Schopenhauer relied on—is critical to understanding Schopenhauer's metaphysics and resolving its apparent contradictions.

Henceforth, I shall use the terms 'meta-consciousness,' 'self-reflection' and 'conscious meta-cognition' interchangeably.

## Chapter 6

# Meta-consciousness in Schopenhauer's metaphysics

*If I were a tree among trees, a cat among animals, this life would have a meaning or rather this problem would not arise, for I should belong to this world. I should be this world to which I am now opposed by my whole consciousness and my whole insistence upon familiarity. ... And what constitutes the basis of that conflict, of that break between the world and my mind, but the awareness of it?*
Albert Camus, in *The Myth of Sisyphus* (1942)

It could be argued that the notion of meta-consciousness discussed in the previous chapter is a discovery of modern psychology and analytic philosophy, unavailable to nineteenth century thinkers. As such—or so the argument may go— Schopenhauer couldn't have relied on it. Yet, contemporaries of Schopenhauer were demonstrably well acquainted with the notion. Indeed, in 1849 Danish philosopher Søren Kierkegaard went as far as to *define* humanness on the basis of our capacity to become self-reflectively aware of our own mentation:

Man is spirit. But what is spirit? Spirit is the self. But what is the self? The self is a relation which relates itself to its own self (Kierkegaard 2013: 269)

This capacity for self-reflection enables what Kierkegaard calls the 'sickness of despair,' a human being's inherent potential to separate itself (as subject) from the world (as object) and then amplify the resulting psychological discomfort through reflective brooding and rumination. Nonetheless, Kierkegaard does acknowledge that self-reflection is *also* our key asset:

The possibility of this sickness is man's advantage over the beast, and this advantage distinguishes him far more essentially than the erect posture, for it implies the infinite erectness or loftiness of being spirit. (*Ibid.*: 272)

In other words, self-reflection is what most essentially distinguishes human beings from other animals, allowing us to embody and achieve something beyond the reach of the rest of nature.

Also fluent in the notion of meta-consciousness was Schopenhauer's one-time enthusiastic disciple, Friedrich Nietzsche. Consider, for instance, this passage from *The Gay Science*:

The problem of consciousness (or *more correctly*: of *becoming conscious of oneself*) meets us only when we begin to perceive in what measure we could dispense with it ... For we could in fact think, feel, will, and recollect ... and nevertheless nothing of it all need necessarily "come into consciousness" ... The whole of life would be possible without it as it were *seeing itself in a mirror*: as in fact even at present the far greater part of our life still goes on without this *mirroring*. (Nietzsche & Common 2006: 166-167, emphasis added)

So for Nietzsche 'consciousness' entails becoming conscious of oneself.[10] It requires the mirroring of re-representation, the explicit awareness of one's own experiential states. That the far greater part of our life goes on without this mirroring means merely that most of our experiences unfold outside the field of self-reflection, as they in fact do. But who would argue that what we think, feel, will and recollect aren't experiential states?

Nietzsche's intended meaning becomes even clearer when he, shortly thereafter, links 'consciousness' to our ability to introspect and report our experiential states, which is a hallmark of meta-consciousness:

> *consciousness generally has only been developed under the pressure of the necessity for communication* ... as the most endangered animal [man] *needed* help and protection; he needed his fellows, he was obliged to express his distress, he had to know how to make himself understood—and for all this he needed "consciousness" first of all: he had to "know" himself what he lacked, to "know" how he felt, and to "know" what he thought ... man, like every living creature, thinks unceasingly; but does not know it; the thinking which is becoming *conscious of itself* is only the smallest part thereof. (Nietzsche & Common, 2006: 167-168, original emphasis)

Clearly, what Nietzsche calls 'consciousness' is actually *meta-consciousness*. Indeed, could there be a more cogent and unambiguous description of the latter than the passage above? Moreover, conflating meta-consciousness with consciousness is, as we've seen, a gaffe common to this day.

Nietzsche and Kierkegaard were not alone in their grasp of meta-consciousness. Even a cursory read of Schopenhauer himself demonstrates unambiguously that he, too, had a clear understanding of the notion:

> Chains of abstract reasoning ... serve to fix the immediate knowledge of the understanding for the faculty of reason ... [so] to be in a position *to point it out and explain it to others.* (W1: 21-22, emphasis added)

He gets even more explicit and precise:

> an entirely new consciousness has arisen, which with very appropriate and significant accuracy is called *reflection. For*

> it is in fact a *reflected appearance,* a thing derived from this
> knowledge of perception (W1: 36, emphasis added)

This reflected appearance is a conceptual echo in consciousness of a lower-level content of consciousness, achieved when consciousness turns in upon itself so to re-represent its own experiences. This is significant also at the level of the single *individual,* independently of his ability to communicate with others, for "By reflection this individual can *make clear to himself* what has been apprehended" (W2: 81, emphasis added).

And now the clincher:

> Reflection is necessarily the *copy* or *repetition* of the originally
> presented world of perceptions ... Concepts, therefore, can
> quite appropriately be called *representations of representations.*
> (W1: 40, emphasis added)

Or, of course, *re-representations!* So what Schopenhauer calls 'abstract representations' or 'conceptual thinking' entails meta-consciousness. He goes on to explain that the process is recursive: one can have re-representations of re-representations of re-representations, etc. There is, thus, a scale of levels of meta-consciousness.

Crucially, Schopenhauer does *not* restrict consciousness to abstract representations:

> We attribute to [all animals] consciousness, and although the
> name (*Bewusstsein*) is derived from *wissen* (to know rationally),
> *the concept of consciousness coincides with that of representation
> in general, of whatever kind it may be.* (W1: 51, emphasis added)

So perception and the intellect, even when unaccompanied by abstract representations, are already *conscious.* Animals have conscious experiences even though they lack abstract representations.

Nonetheless, what Schopenhauer means by the word 'consciousness' does already intrinsically involve some subtle level of self-reflective cognitive processing done by the intellect, even though less than abstract representations. In other words, for Schopenhauer there is more to 'consciousness' than mere 'what-it-is-likeness' or raw experiential states:

> self-consciousness ... contains a knower and a known, *otherwise it would not be a consciousness.* (W2: 202, emphasis added)[11]

So 'consciousness' requires the knower-known pair characteristic of self-reflection, whereas mere raw experience or 'what-it-is-likeness' doesn't. Schopenhauer is again explicit in this regard when he states, "representation as such already *presupposes* a form, namely *object and subject*" (W1: 162, emphasis added). This applies to 'consciousness' as well, since Schopenhauer equates it with representation.

It is because of this restrictive attribution of self-reflection to 'consciousness' — which echoes Nietzsche, Jung, Cleeremans, Dijksterhuis, etc. — that Schopenhauer can say:

> The will in itself is *without consciousness,* and in the greatest part of its phenomena remains so. The secondary world of representation must be added for the will to become *conscious of itself,* just as light becomes visible only through the bodies that *reflect* it (W2: 277, emphasis added)

Notice that here he again associates 'consciousness' with reflection. At one point, when discussing plants as a "symbol of consciousness" (W2: 202-203), Schopenhauer effectively *defines* 'consciousness' as entailing the self-reflective knower-known pair.

I submit that, according to Schopenhauer, the know*er* in any kind of representation, which is associated with our feeling of ipseity or I-ness, is merely a cognitive mirror that reflects—i.e. re-represents—lower-level, more primary experiential states. As a matter of fact, he says as much:

> in self-consciousness the know*n*, consequently the will, must be the first and original thing; the know*er*, on the other hand, must be only the secondary thing, that which has been added, *the mirror*. They are related somewhat as the self-luminous is to the *reflecting* body; or as the vibrating strings are to the sounding-board, *where the resulting note would be consciousness*. (W2: 202, emphasis added)

Once again, 'consciousness' is the reflection, the re-representation. Even in mere perceptual representations, there is already a know*er*—a cognitive mirror—re-representing the immediate experiential states of sense impression, which in turn become its know*n*. Perceptual representations thus already entail a subtle level of self-reflection.

To sum it up, although all instances of abstract representations are re-representations, not all instances of re-representations in Schopenhauer's metaphysics are abstract representations. The latter presuppose, as their ground, fairly high-level mental functions that, by themselves, already encompass re-representations. Moreover, re-representations of endogenous, non-sensory *feelings*—such as those of "hatred, disgust, self-satisfaction, honour, disgrace, right and wrong," etc. (W1: 51)—are not abstract representations, for the latter must be grounded in perceptual representations whereas endogenous feelings aren't representations at all.

The notion of meta-consciousness pervades Schopenhauer's metaphysics. Not only is it explicitly present in the form of abstract representations, it is also more implicitly present in representations in general. Schopenhauer seems to assume a continuum of levels of meta-consciousness, ranging from zero in inanimate nature—i.e. the primordial or 'blind' configuration of the will—to subtle levels in perceptual representations, to higher levels in concepts grounded in perception, to even higher levels in abstract representations of abstract representations.

And here we can already elucidate what may be one of the most noticeable seeming contradictions in Schopenhauer's metaphysics: his inconsistent, context-dependent use of the word 'will.'

Indeed, according to Schopenhauer the will is the inner essence of *everything* in nature, without exception. There is *nothing* that isn't will.[12] Yet, he also often uses the word 'will' in a more restrictive sense: for instance, when discussing the relationship between intellect and will as if they were separate entities (W2: 208-210). He describes the will as "the strong blind man carrying the sighted lame man [i.e. the intellect] on his shoulders" (W2: 209). But if the will is the in-itself of everything, then surely there is nothing to the intellect but will. How can Schopenhauer then talk of the two as though they were different?

As the sighted lame man, the intellect is merely a *local, superficial configuration* of the will—resulting from a segment of the latter turning in upon itself—by virtue of which the will achieves some degree of self-reflection in the form of re-representation. In other words, the intellect is a local meta-cognitive structure of and in the will. Its extrinsic appearance or representation is what we call a brain. The intellect is lame compared to the broader will because it is but a small, superficial segment of the latter, a tiny beacon of reflection sitting atop a mountain of instinct. Yet the will—the strong man—is blind without the lame man, because it then lacks the 'sight' of meta-conscious representation.

This way, what Schopenhauer means by 'will' is often not the will as a whole, but the *raw, non-meta-conscious layer of the will underlying the intellect*; a layer still in the will's original, purely instinctive configuration. In other words, he contrasts the intellect with what is left of the will *without* the intellect; the tiny beacon with the mountain. That he often loosely refers to the latter simply as 'will' is just shorthand for 'instinctive underlying layer of the will'; it doesn't actually contradict his contention that there ultimately is nothing to the intellect but the will *proper*, now in the full sense of the word.

Such mild terminology inconsistencies—entirely benign and hardly worthy of note in a colloquial setting—pervade Schopenhauer's argument. It is nonetheless clear throughout that, ultimately, there is only the will. Even representation is merely the objectification or extrinsic appearance of the will. The latter can configure itself—by locally turning in upon itself—according to a continuum of levels of meta-cognition. In inanimate nature this level is zero, and so there we have the will purely in its primordial, raw, blind configuration. This primordial will—lacking meta-cognition and therefore incapable of differentiating itself, as the recipient of experience, from its own experiential contents—is what Schopenhauer makes a point to contrast with the 'intellect' and 'reason.'

## Chapter 7

# The will's strife for meta-consciousness

*Blood, torment and sacrifice were necessary for man to create*
*memory in himself ... With the help of this kind of memory man*
*eventually acquired 'reason'! Ah, reason, solemnity, mastery over*
*the emotions, this whole gloomy, dismal thing which is called*
*reflection, ... how dear the price they have exacted!*
Friedrich Nietzsche, in *On the Genealogy of Morals* (1887)

Schopenhauer speaks of (a) the blindly acting forces of inanimate
nature, (b) plants' spontaneous reactions to stimuli such as
sunlight, (c) human motives for action, and even (d) our most
sophisticated abstract reasoning, as *degrees of manifestation of the*
*same underlying will:*

> [The will] appears in every blindly acting force of nature, and
> also in the deliberate conduct of man, and the great difference
> between the two concerns only *the degree of manifestation*, not
> the inner nature of what is manifested. (W1: 110, emphasis
> added)

For him, nature consists of a continuum of manifestations of
the will, which *spontaneously*—thus *not* deliberately—strive
toward a developmental climax (W1: 144-152). I submit that
this developmental climax is the achievement of higher levels of
*meta-consciousness*. Allow me to elaborate.

According to Schopenhauer, corresponding to each degree
of manifestation of the will in nature there is a "grade of
objectification" (W1: 128-129). The degree of manifestation
applies to the thing in itself, whereas the grade of objectification

applies to the corresponding appearance or representation. In our case, for instance, the grade of objectification of the will has to do with the visible forms of our body, whereas its degree of manifestation has to do with what it is like to *be* us as self-reflective conscious beings. Schopenhauer writes:

> The most universal forces of nature [e.g. gravity, electromagnetism] exhibit themselves as the lowest grade of the will's objectification ... they are *immediate* phenomena of the will (W1: 130, emphasis added)

The qualifier 'immediate' denotes lack of meta-consciousness. Schopenhauer also describes this lowest degree of manifestation of the will as "a *blind* impulse" (W1: 149, emphasis added), which emphasizes the same point. He then proceeds:

> Objectifying itself more distinctly from grade to grade, yet still completely without *knowledge* as an obscure driving force, the will acts in the plant kingdom. (W1: 149, emphasis added)

While contending that plants embody a *higher* degree of manifestation of the will than the universal forces, Schopenhauer insists that they, too, lack 'knowledge'—i.e. representation. In what sense, then, do plants embody this higher degree? It can *only* be the plants' ability to *apprehend and react to stimuli* (W1: 115-116), such as when a sunflower turns to follow the sun. And like perceptions, stimuli also provide a portal to the external environment. So why doesn't Schopenhauer consider the apprehension of stimuli a—perhaps lower, less sophisticated—form of *perceptual representation*?

The answer is two-fold: first, unlike perceptions, stimuli are not accompanied by the spontaneous intellectual process that identifies cause-and-effect relationships in sense data.

Second—and most importantly—since Schopenhauer attributes a subtle level of self-reflection to perceptual representations, distinguishing stimuli from perceptions avoids the implausible implication that plants self-reflect, even though they supposedly experience raw qualia.[13] Indeed, the distinction circumvents what would otherwise be an implicit suggestion that a plant has a sense of self—i.e. of being an individual subject separate from an objectified world that stimulates it.

Schopenhauer insists that, unlike representations, stimuli do not entail self-reflection even when they occur in the human body:

> Even in us the same will in many ways acts *blindly*; as in ... all [of the body's] vital and vegetative processes, digestion, circulation, secretion, growth, and reproduction. ... here this will is *not guided by knowledge* [i.e. representation] ... not determined according to motives, but acts *blindly* according to causes, called in this case *stimuli*. (W1: 115, emphasis added)

Unlike our deliberate actions, we cannot—at least ordinarily—re-represent the volitional states underlying our autonomic functions. Triggered by stimuli, these functions—though still experiential in essence—unfold below the threshold of meta-conscious introspection.

Moving on to the next grade:

> The higher and higher grades of the will's objectivity lead ultimately to the point where ... movement consequent on motives and, because of this, *knowledge*, ... becomes necessary ... the world as *representation* now stands out at one stroke (W1: 150, emphasis added)

These are thus animals capable of experiencing the world

through representation. But

> Animals ... have merely representations from perception, no concepts (W1: 151)

Animals are not yet as meta-conscious as humans, which embody the highest degree of manifestation of the will:

> A higher power of knowledge of perception, so to speak, had to be added to this, a *reflection* of that knowledge of perception ... With this there came into existence ... *deliberation* ... and finally the fully distinct *consciousness of the decisions of one's own will as such*. (W1: 151, emphasis added)

Here it is clear that what Schopenhauer sees as the highest degree of manifestation of the will is *full conscious self-reflection*. He adds:

> man is nature ... at the highest grade of her *self-consciousness* (W1: 276, emphasis added)

Recursive meta-consciousness is thus the developmental climax that the will is 'blindly' striving toward through a series of degrees of manifestation, namely:

1. The latent presence of 'blind' universal forces throughout the fabric of nature.
2. The rise of organisms capable of apprehending and reacting to *stimuli*, an in-between step in the direction of representation.
3. The rise of organisms capable of *perceptual representations*, more complex forms of stimulus apprehension that entail cognition, subtle levels of self-reflection and can, in principle, be re-represented.

4. The rise of organisms capable of actually *re-representing* perceptual representations.

5. The rise of organisms capable of *re-representing re-representations*, in a recursive manner.

Each degree in this sequence is a step toward the achievement of recursive meta-consciousness, the

> victory of the Idea of the organism, *conscious of itself*, over the physical and chemical laws [i.e. blind universal forces] which originally controlled the humours of the body. (W1: 146, emphasis added)

It should be emphasized that it is the same will that underlies all five steps in the sequence above (W2: 318). In its primordial state in step 1—which Schopenhauer calls the "will-without-knowledge that is the foundation of the reality of things" (W2: 269)—the will is 'blind,' in the sense of having no self-reflection. Steps 2 to 5 correspond then to local *reconfigurations* of the will whereby it progressively turns in upon itself so to reflect or mirror its own experiential states. Schopenhauer neatly summarizes all this in the opening paragraph of Chapter 25 of the second volume of *The World as Will and Representation* (W2: 318).

## Chapter 8

# Resolving the key seeming contradiction

*Communicative meaning is first incarnate in the gestures by which the body spontaneously expresses feelings ... The gesture is spontaneous and immediate. It is not an arbitrary sign that we mentally attach to a particular emotion or feeling; rather, the gesture is the bodying-forth of that emotion into the world, it is that feeling of delight or of anguish in its tangible, visible aspect.*
David Abram, in *The Spell of the Sensuous* (1996)

The understanding that meta-consciousness is the basis of the degrees of manifestation of the will enables us to resolve the key apparent paradox in Schopenhauer's metaphysics: the tension between the claims that (a) the will cannot be known, for it is rooted outside spacetime; and that (b) we *can* know the will through our first-person experience—the intrinsic view or concealed order—of ourselves.

The key is to realize that what Schopenhauer means by 'to know the will' is a slightly different thing in each case. In claim (a), the will cannot be known insofar as it cannot be *fully re-represented*. When we don't know *that* we know an aspect of the will, we cannot report our knowledge of it even to ourselves, so everything unfolds *as if* we truly didn't know it. In claim (b), on the other hand, the will *can* be known in the sense that it can be *experienced—felt—in an immediate manner*, even if such experience is not re-represented.

Importantly, Schopenhauer opens the door to a grey area:

This will constitutes what is most immediate in [man's] consciousness, but as such *it has not wholly entered into the form of representation* ... it makes itself known in an immediate way in which *subject and object are not quite clearly distinguished* (W1: 109, emphasis added)

The implication here is that the will can *partly* enter spacetime — for representation unfolds across the individual subject's cognitive spatio-temporal scaffolding, so to partly enter into the form of representation implies partly entering spacetime — and, consequently, be partly *re*-represented. But because this re-representation is then itself also only *partial*, the will can be known only in a manner in which subject and object are not quite distinct: the associated level of self-reflection isn't zero, but also isn't sufficient to equip the individual subject with a clear sense of a self separate from its object. In conclusion, *the will can be re-represented just enough for us to get a vague feeling of it through introspection*, a hint of its existence and nature.[14]

And this is the best we can hope for, for according to Schopenhauer the split of the will into the two halves of know*er* and know*n* — which is intrinsic to meta-consciousness even according to Schopenhauer himself[15] — hampers our knowledge of it, insofar as the half that knows cannot then itself be known (W2: 196-197). Ideally, the will would become explicitly aware of itself without splitting itself into these two halves, but this is of course a contradiction in terms.

Notice that, because Schopenhauer *defines* abstract representations as meta-cognitive reflections of either lower-level abstract representations or perceptual representations, all abstract representations are ultimately grounded in perception. Therefore, the will — which is *prior* to, and *independent* of, perception — cannot be characterized through abstract representations.

*But this doesn't mean that the will can't be re-represented.* To see how, notice first that feelings differ from representation in that they precede and can partly bypass the operation of the intellect:

> The changes experienced by every animal body are *immediately* known, that is to say, *felt*; and as this effect is referred at once to its cause, there arises perception of the latter as an object. (W1: 11-12, emphasis added)

Without this reference to a cause, done by the intellect,

> perception would never be attained; there would remain only a dull, *plant-like consciousness*[16] of the changes (W1: 12, emphasis added)

That is, conscious *feelings*.

But *endogenous* conscious feelings unmediated by the sense organs—such as inner yearnings, longings, cravings, anxiety, dread, etc.—can obviously *still be re-represented*. We do it all the time—i.e. we access our inner feelings through introspection—independently of perception.

I submit that this is how the "immediate affections of the will" (W1: 101) can be partly re-represented and explicitly known: we *feel* the will as subtle, endogenous, volitional experiential states, without having clarity regarding whether we *know* or *are* these feelings—i.e. without clear distinction between know*er* and know*n*, subject and object. Schopenhauer says basically as much when he states that we wouldn't be able to understand the character of livings beings "if the inner essence of things [i.e. the will] were not otherwise known to us, at least obscurely and *in feeling*" (W2: 364, emphasis added).

My contention is summarized in the scheme of Figure 1. On the top part of the figure, the cognitive chain between stimuli and abstract representations is shown. Schopenhauer is quite explicit in this regard, so I will elaborate on this merely for the sake of completeness: first, a stimulus impinges on an organism's sense

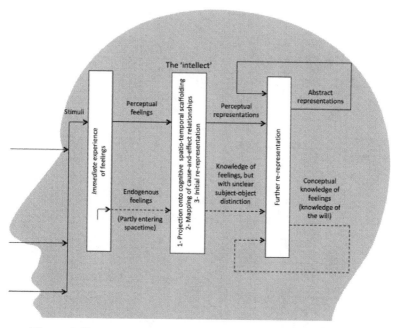

**Figure 1.** The conscious processing of stimuli and endogenous feelings, according to Schopenhauer.

organs, which produces the *immediate* experience of a feeling. This immediate experience is then at once referred to the intellect, where, as we have seen, three processing steps take place: first, the immediate experience is projected onto the organism's internal cognitive spatio-temporal scaffolding; second, a network of cause-and-effect relationships is spontaneously mapped onto the projected experiences; third, an initial, subtle re-representation takes place, so the individual subject distinguishes itself, as know*er*, from its know*n* object. The output of these three steps of cognitive processing are perceptual representations, which can then be recursively further re-represented for the generation of abstract representations.

However, as Schopenhauer insists, the will is known not through stimuli, but as *endogenous* feelings arising entirely within the individual subject (indeed, this is precisely what distinguishes the will from representation). These endogenous feelings can only *partly* be referred to the intellect—which is indicated by dashed lines in the figure—because the intellect is itself a manifestation of the will and cannot, as such, process itself. Once in the intellect, the referred part of the endogenous feelings is projected onto the organism's cognitive spatio-temporal scaffolding and becomes known in a manner in which "subject and object are not quite clearly distinguished" (W1: 109). This partial knowledge can nonetheless be further re-represented, so that the original endogenous feelings can be partly conceptualized and accessed in rational thought (W2: 209-210). This is the manner in which the will—manifesting itself *as* the endogenous feelings—can be known and talked about in an explicit sense, even though the endogenous feelings themselves remain always distinct from representations of any kind, as they are not grounded in stimuli. Indeed, I contend that this is what Schopenhauer means when he writes that the will "can be known by [the individual subject] only indirectly, through reflection as it were" (W2: 278). Such indirect but explicit knowledge does not exclude the earlier, direct but implicit knowledge attained with the *immediate* experience of the endogenous feelings.

Although Schopenhauer is admittedly less clear about the process illustrated in the bottom part of Figure 1 than that in the top part, I submit that there is enough in his words to allow me to attribute the complete scheme in Figure 1 to Schopenhauer himself.

The discussion above reveals the origin of the seeming paradox in Schopenhauer's claims about our ability—or lack thereof—to know the will. There is no actual contradiction in these claims, as the spirit of Schopenhauer's argument is entirely coherent in the context of a clear understanding of meta-consciousness.

# Chapter 9

# Schopenhauer and quantum mechanics

*[A]lthough physicists talk of atoms and other microscopic entities as if they were physical things, microscopic things are only concepts we use to describe the behavior of our measuring instruments. ... The quantum theory of microscopic objects should explain the sensible behavior of our macroscopic equipment, but microscopic objects themselves need not "make sense." Consider an analogy from psychology ... We report on a person's behavior. The physical behavior itself presents no paradox. ... A person's motives, however, are theories that should explain the person's behavior. But the motives themselves need not, and often do not, make sense.*
Bruce Rosenblum and Fred Kuttner, in *Quantum Enigma* (2006)

We've seen thus far that, for Schopenhauer, the world out there, as it is in itself, is (a) constituted by experiences of a volitional nature and (b) fundamentally unitary, outside spacetime. These two statements reinforce each other's plausibility: whereas it's arguably impossible to explicitly conceive of a *physical* world outside spacetime, extra-spatiotemporal experiential states have been consistently reported in mystical literature throughout history, and are still routinely reported today by meditators and psychonauts alike. Moreover, whereas multiple volitional experiences can, in fact, be just nominal facets of an ultimately unitary experiential gestalt—think of the experience of feeling desire, hope and fear simultaneously, in an overlapping manner— definite physical objects or events cannot occupy the same volume of space at the same time. Therefore, Schopenhauer's characterization of the world-in-itself is internally consistent.

In addition, (c) what we call the 'physical world' of definite objects exists, for Schopenhauer, purely in representation and is, therefore, relative to the individual subject who observes it: "The whole world of objects is ... wholly and for ever conditioned by the subject" (W1: 14-15). This statement (c) coheres with statements (a) and (b) above: by positing that separate physical objects and events do exist, *but only in the mind of the individual observer*, Schopenhauer accounts for our empirical experience of the physical world without contradicting statements (a) and (b).

But here I want to go beyond acknowledging the internal consistency of Schopenhauer's view of the world; I want to explore whether our modern scientific understanding of nature contradicts or, in fact, *supports* Schopenhauer's views. To do so, we must review salient developments in Quantum Mechanics (QM) over the past few decades.

According to popular intuition, the physical properties of an object should exist and have definite values even when the object is not being observed. For instance, the moon should exist and have whatever weight, shape, size, color, position and movement it has even when nobody is looking at it. Moreover, a mere act of observation should not change the values of these properties.

Operationally, this intuition is formalized in the notion of 'non-contextuality': the outcome of an observation should not depend on the way another, separate but *simultaneous*, observation is performed. For instance, consider the situation depicted in Figure 2: when a small light source is placed in a fiber optics cable, shooting different photons *A* and *B* towards either end of the cable, what Alice sees when she looks at photon *A* through instrumentation should not depend on how Bob simultaneously observes photon *B*. In other words, the observed properties of photon *A* should not depend on the simultaneously observed properties of photon *B*.

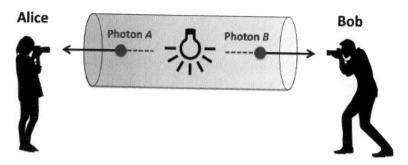

**Figure 2.** If photons *A* and *B* existed as definite and separate objects, what Alice sees when she measures photon *A* shouldn't depend on how Bob simultaneously measures photon *B*. However, quantum physics has shown that such is not the case: photons *A* and *B* are inherently intertwined; they cannot be treated as separate and definite objects.

Such non-contextuality, if true, could be construed to undermine Schopenhauer's metaphysics: it would substantiate the notion that objects—such as photons—exist in a definite and individuated state *even outside representation*; after all, photon *A* would seem to continue to be whatever it is, regardless of how photon *B* presents itself to observation.

However, according to QM, the outcome of an observation *can* depend on the way another, separate but *simultaneous*, observation is performed: if the photons are generated in a special way, what Alice sees when she observes her end of the cable *does* depend on how Bob simultaneously observes the photon coming out of his end; even if the cable is arbitrarily long. Take a moment to absorb this. If each photon were a definite, separate object with physical properties independent of observation, how could this so-called 'entanglement' be possible? How could an observation of something on one end of a cable *instantaneously* influence the observation of something *else* on the *other* end?

Hoping to make sense of this, Albert Einstein posited that such 'spooky action at a distance,' as he put it, must be merely an artifact of an alleged incompleteness of QM: each photon must have 'hidden' properties missed by QM, which somehow account for the correlations between the observations on the two ends of the cable (Einstein, Podolsky & Rosen 1935). Alas, Einstein was proven wrong when physicist John Bell demonstrated, mathematically, that the predictions of QM cannot be accounted for by any hypothetical hidden property of the photons (1964). Ergo, if QM is correct, non-contextuality cannot be true.[17] So is QM correct?

Since Alain Aspect's seminal experiments (Aspect, Grangier & Roger 1981, Aspect, Dalibard & Roger 1982, Aspect, Grangier & Roger 1982), the predictions of QM in this regard have been repeatedly confirmed. The year 1998 was particularly fruitful, with two remarkable experiments performed in Switzerland (Tittel *et al.*) and Austria (Weihs *et al.*). More recent experiments again challenged non-contextuality (Lapkiewicz *et al.* 2011, Manning *et al.* 2015). Commenting on them, physicist Anton Zeilinger has been quoted as saying that "there is no sense in assuming that what we do not measure [that is, observe] about a system has [an independent] reality" (Ananthaswamy 2011). Finally, Dutch researchers (Hensen *et al.* 2015) and a large international collaboration (The BIG Bell Test Collaboration 2018) successfully performed tests closing all potential loopholes and definitively proving QM correct.

The only alternative left for those holding on to non-contextuality is to postulate some form of non-locality: nature must have—or so they speculate—hidden physical properties that are somehow 'smeared out' across spacetime and connect the two ends of the fiber optics cable instantaneously. It is this imaginary, omnipresent, invisible but allegedly *observation-independent* background that supposedly orchestrates, from behind the scenes, quantum entanglement.

It turns out, however, that some predictions of QM are

incompatible with non-contextuality *even for a large and important class of non-local theories* (Leggett 2003). Experimental results have now confirmed these predictions (Gröblacher *et al.* 2007, Romero *et al.* 2010). To reconcile such results with popular intuition would require an arbitrary redefinition of what we call 'physical objectivity.' And since contemporary culture has come to conflate physical objectivity with reality itself, the science press felt compelled to report on these experiments by pronouncing, "Quantum physics says goodbye to reality" (Cartwright 2007).

The surviving non-local theory that could, at least in principle, still preserve non-contextuality is 'De Broglie-Bohm's Pilot Wave Theory'—or, more simply, 'Bohmian Mechanics' (Bohm 1952a, Bohm 1952b). Alas, this theory is plagued by a number of problems. For instance, unlike regular QM with its Quantum Field Theory extensions, Bohmian Mechanics has no relativistic version that could be reconciled with Einstein's Relativity. Physicists Raymond Streater (2007: 103-112) and Luboš Motl (2009) reviewed other technical arguments against Bohmian Mechanics in their work. Even the theory's own originator, Louis de Broglie, concluded against it after David Bohm completed the framework in the 1950s.

Admittedly, there is still polemic surrounding not only Bohmian Mechanics, but also the experimental results that seem to refute non-contextuality. It is nonetheless fair to say that, never before in the history of Western thought, at least since the Enlightenment, has the idea of a definite physical world independent of observation looked so precarious. Non-contextuality, if not dead, is in life support.

And here are the key implications, as far as Schopenhauer's metaphysics is concerned: without non-contextuality, there are (a) no physical objects or events, with definite properties and occupying definite positions in spacetime, before observation. The world as it is in itself, independently of observation, exists instead in a so-called 'quantum superposition': a state of overlapping potentialities or tendencies, but no concrete 'physicality' in the sense we ordinarily attribute to the word. There is also (b) no tenable way to carve out separate objects and events in the world-in-itself. After all, prior to observation everything in the world remains quantum-mechanically entangled and, as Jonathan Schaffer observed, "there is ... good reason to treat entangled systems as *irreducible wholes*" (2010: 32, emphasis added).

These implications of the experimental refutation of non-contextuality confirm Schopenhauer's claims about the unitary, non-physical nature of the world outside representation.

But let us remain cautious: what is meant by 'observation' in physics is a *measurement*, something arguably broader than Schopenhauer's perceptual representations. A perceptual representation certainly entails a measurement, but many would argue that measurements can also be performed—for instance, by electronic detectors—without accompanying perceptual representations.

That said, physicists Henry Stapp, Menas Kafatos and myself have argued that, ultimately, only a *conscious observer*, through representing the physical system under observation in his or her consciousness, can perform true measurements (Kastrup, Stapp & Kafatos 2018). For convenience, I shall now briefly summarize our argument.

The claim that *inanimate* objects—such as electronic detectors—can perform quantum mechanical measurements is fundamentally problematic, because the partitioning of the world into discrete inanimate objects is merely nominal to begin with. Is a rock integral to the mountain it helps constitute? If so, does it become a separate object merely by virtue of its getting detached from the mountain? And if so, does it then perform a measurement each time it comes back in contact with the mountain as it bounces down the slope? Brief contemplation of these questions shows that the boundaries of a detector are arbitrary. Indeed, as first argued by John von Neumann (2018) and rearticulated in Stapp's work (2001), when two inanimate objects interact they simply become entangled with one another — that is, they become united in such a way that the behavior of one becomes inextricably linked to the behavior of the other — but no actual measurement is performed. As such, the *inanimate* world is a unitary, indivisible physical system governed by QM. There are no detectors performing measurements; there is only the one inanimate world.

Let me use a concrete example so to be more specific. In the well-known double-slit experiment, electrons are shot through two tiny slits. When they are observed at the slits, the electrons behave as definite individual particles. But when observed only *after* they've passed through the slits, the 'electrons' behave as *superposed potentialities*. In 1998, researchers at the Weizmann Institute in Israel showed that, when detectors are placed at the slits, the electrons behave as definite individual particles (Buks *et al.*). At first sight, this may seem to indicate that measurement does not require a conscious observer.

However, *the output of the detectors only becomes known when it is consciously observed by a person*. The hypothesis of a measurement before this conscious observation lacks compelling theoretical or empirical grounding. After all, QM offers no reason why the whole system — electrons, slits and detectors combined — shouldn't be in an entangled superposition before and until someone looks at the detectors' output (von Neumann 2018). How are we to know? Since we cannot abstract ourselves out of our knowledge, we cannot know that detectors actually perform measurements.

Consequently, as far as we *can* know, before it is *represented* — through conscious perception — the world consists of a unitary superposition of potentialities or tendencies. This superposition — indivisible, as quantum entanglement prevents elements of the superposition from being describable separately from one another—is incompatible with the existence of individual, separate objects or events with definite properties. Again, *this seems to confirm Schopenhauer's view of the world outside representation.*

Moreover, insofar as concrete 'physicality'—in the sense we ordinarily attribute to the word—comes into being only through an act of conscious observation, its existence is tied to perceptual representation and, therefore, relative to the observing individual subject. I've made an empirical case for this above. But there is also a compelling theoretical case, whose implications for our understanding of the importance and prescience of Schopenhauer's metaphysics cannot be underestimated.

Indeed, if we stick to plain quantum theory and forget about imaginary hidden properties—be they local or global—what does QM tell us about the world? Physicist Carlo Rovelli answered this question rigorously (1996) and the result is now known as Relational Quantum Mechanics, or simply 'RQM.' According to RQM, there are no absolute—that is, observer-*in*dependent—physical quantities. Instead, *all* physical quantities—the entire physical world—are *relative* to the observer in a way analogous to motion, a significant point that has now been *experimentally confirmed* (Proietti *et al.* 2019; see also Emerging Technology from the arXiv 2019). Consequently, each observer is inferred to 'inhabit' his or her own physical world, as defined by the context of his or her own observations.

Rovelli summarizes RQM thus:

[Because] different observers give different accounts of the same sequence of events, ... each quantum mechanical description has to be understood as relative to a particular observer. Thus, a quantum mechanical description of a certain system (state and/or values of physical quantities) cannot be taken as an "absolute" (observer-independent) description of reality, but rather as a formalization, or codification, of properties of a system relative to a given observer. (1996: 1648)

This echoes with uncanny precision the "complete and universal *relativity* of the world as representation" (W1: 34, emphasis added) claimed by Schopenhauer. It also circumvents the experimental paradoxes posed by the evidence against non-contextuality: if there is no absolute physical world shared by us all to begin with, of course there are no definite objects out there waiting to be observed. Moreover, the discussion about 'spooky action at a distance' becomes meaningless: the motivation for it was that Bob's actions on his end of the fiber optics cable seem to *instantaneously* influence what happens on Alice's end (see Figure 2 again). But this only suggests non-locality under the assumption that both Alice and Bob operate in *the same physical world:* something non-local in their shared physical environment should then coordinate their respective observations across spacetime. According to RQM, however, Alice and Bob do *not* share the same physical world to begin with, and so the whole issue of non-locality melts away.

It's not all roses, though: RQM's elegant escape from experimental paradoxes comes at the cost of a number of metaphysical qualms. First, the idea that the physical world one inhabits is a product of one's own observations seems to imply solipsism, an anathema in philosophy. Second, RQM posits that "a complete description of the world is exhausted by the relevant [Shannon] information that systems have about each other" (Rovelli 1996: 1650). However, according to Claude Shannon (1948), information isn't a thing unto itself; it is merely a way to quantify the possible discernible states of a substrate. So if there is no absolute physical substrate, what exactly is it whose possible discernible states are being quantified by information? Third—and perhaps most problematic of all—the RQM tenet that all physical quantities are relative raises an obvious question: *relative to what?* We only recognize meaning in a relative quantity, such as e.g. motion, because we assume there to be *absolute* physical bodies that move with respect to one another. But RQM denies *all* physical absolutes that could ground the meaning of relative quantities.

Notice that the root of all these metaphysical qualms is the unexamined assumption that *only physical quantities exist.* If physical quantities arise from personal observation and they are all there is, then solipsism is indeed implied. If physical quantities are grounded in information and they are all there is, then information indeed lacks a substrate. If physical quantities are relative and they are all there is, then there are indeed no absolutes to ground their meaning.

By identifying this hidden—yet far-reaching—assumption we can come to a remarkable realization: *Schopenhauer already laid the groundwork for resolving the metaphysical qualms of RQM more than 200 years ago.* Allow me to elaborate.

Stanford physicist Andrei Linde, of cosmic inflation fame, once rendered the epistemic scope of physics painfully explicit:

> Let us remember that our knowledge of the world begins not with matter but with perceptions. I know for sure that my pain exists, my 'green' exists, and my 'sweet' exists ... everything else is a theory. Later we find out that our perceptions obey some laws, which can be most conveniently formulated if we assume that there is some underlying reality beyond our perceptions. This model of material world obeying laws of physics is so successful that soon we forget about our starting point and say that matter is the only reality, and perceptions are only helpful for its description.[18]

Hence, because it excludes any absolute, observer-independent physical substrate—i.e. the "underlying reality beyond our perceptions" or the "material world obeying laws of physics"—the physical world of RQM *can only be the contents of perception.* There is just nothing else for it to be. Take a moment to convince yourself of this counterintuitive but unavoidable implication.

The unexamined assumption underlying the metaphysical qualms of RQM—namely, that only physical quantities exist—can then be restated as follows: *only the contents of perception exist.* Such an alternative but equivalent formulation renders the untenability of the assumption patently clear: next to the contents of perception there are, of course, also non-perceptual experiential states, *such as volition.* In Schopenhauer's terminology, *next to representation there is also the will!*

Many physicists posit that volitional states should be explainable in terms of physical quantities and, as such, become part of the physical world by reduction. But this, too, is a metaphysical *assumption* that does not change the scientific *fact* that quantum mechanics does *not* predict volitional states; it only predicts the unfolding of perception, even when what is predicted—and later perceived—is the output of instrumentation.

Schopenhauer thus sketches—*avant la lettre*—a solution to the metaphysical qualms of RQM by positing that the world-in-itself is not constituted by physical quantities, but by the volitional states of the will. These states are the absolutes that ground all physical quantities: the latter arise on the screen of perception as *relationships* between volitional states (more on this in the next chapter). The will also provides a substrate for what we call 'information': the concept, which is so fundamental to RQM, reflects but possible discernible states *of the will.* Finally, a universal will—in which we, as individual subjects, are all immersed—avoids solipsism, even though each individual observer has his or her own physical world.

In a nutshell, *the will is the missing metaphysical ground required by RQM;* it fits the latter like a glove. Schopenhauer's metaphysics, almost one hundred years before the advent of QM, had already laid the groundwork for the resolution of quantum paradoxes. As a matter of fact, these paradoxes would never have arisen if we, as a culture, had heeded Schopenhauer's metaphysics and allowed it to shape our intuitions.

# Chapter 10

# Individuality and dissociation

*Life has been created quite truthfully in order to surprise us (where it does not terrify us altogether).*
Rainer Maria Rilke, in a letter written in the late 19th or early 20th century

For Schopenhauer, our seemingly *individual* subjectivity is merely an epiphenomenon of the universal will, a form of its manifestation, not a fundamental or primary entity. He writes:

> Death is sleep in which *individuality is forgotten*; everything else awakens again, or rather *has remained awake*. (W1: 278, emphasis added)

So individuality is akin to a thought that can simply be forgotten; a transitory experience arising and dissipating in something that always remains awake (i.e. conscious): the universal will itself.

He clarifies that "the individual … does not rest on a self-existing unit" (*Ibid.*)—i.e. the individual doesn't exist in or by itself, in the same way that e.g. a thought doesn't exist in or by itself but is simply a particular manifestation of the underlying mind. Indeed, later on Schopenhauer speaks of individuality as a

> mere condition or *state* [of the will. It has] only a conditioned, in fact, properly speaking, a *merely apparent* reality (W2: 278, emphasis added)

Therefore, for Schopenhauer the existence of multiple individual subjects is an illusion, for "there is only one being" (W2: 321) in nature. Only the unitary, universal will is ultimately real, individual subjects being just something the will *does*. Individuals are experiential *actions* or *behaviors* of the will.

The question that then immediately confronts us is: Just *how* does the will do this? What is the mechanism by means of which it *does* multiple individual subjects concurrently?[19] How is the illusion conjured up?

We know that the experiential states of individual subjects are not integrated across their respective psyches; otherwise we would be able to read each other's thoughts and access each other's memories and feelings. This "disruption of and/or discontinuity in the normal integration" (Black & Grant 2014: 191) of the universal will's experiential states is analogous to what is called *dissociation* in modern psychiatric parlance (American Psychiatric Association 2013). Severe forms of dissociation can cause a person to manifest multiple, seemingly disjoint personalities or centers of consciousness—called 'alters'—in what has become known as Dissociative Identity Disorder, or 'DID.' Recent neuroimaging research has objectively—and compellingly—confirmed the reality of DID (e.g. Schlumpf *et al.* 2014, Strasburger & Waldvogel 2015).

I submit that, implicit in Schopenhauer's argument, is the notion that *individual subjects arise, analogously to alters, as a consequence of DID-like dissociation of the universal will.* After all, as we've seen, the will is consciousness and, as such, can at least in principle undergo dissociation. We can then imagine that, through dissociating itself into multiple, seemingly disjoint centers of consciousness, the will creates the illusion of individual subjectivity, just as DID patients experience multiple seemingly separate personalities in a single mind.

Although the seeds of the modern psychiatric understanding of dissociation were sown by Pierre Janet only after Schopenhauer's death, Schopenhauer himself already offers a cogent description of the role of cognitive *associations* in the ordinary integration of experiential states (W2: 133-136). He explains that, without the thread of integration formed by how *"one* idea draws in *another* by the bond of association" (W2: 137, original emphasis), myriad experiential states would remain forgotten, outside our individual awareness or ego (W2: 140). Having understood this so clearly, it's unlikely that Schopenhauer would have then missed the immediate next step in this line of reasoning: *the dissolution of bonds of association*—i.e. 'dissociation—*can split an otherwise integrated consciousness into seemingly disjoint segments.* Indeed, he proceeds to describe what is effectively a dissociative split between intellect and deeper layers of the will (W2: 208-210). Evocative statements like "occasionally the intellect does not really trust the will" (W2: 210) suggest that Schopenhauer considered the formation of autonomous and co-conscious mental complexes—hallmarks of severe dissociation—an empirical matter of fact.

Clearly, dissociation isn't a foreign notion to Schopenhauer. It's not a great leap, thus, to fill in the gap and posit that, in his metaphysics, dissociation of the will—a form of universal DID—is what accounts for the illusion of individual subjectivity. Schopenhauer carefully assembles all the building blocks necessary for deriving this conclusion, stopping just short of securing it with an explicit claim.

A criticism that could be offered at this point is this: whereas we can perceive and interact directly with other individual subjects in ordinary waking life—after all, I can surely see and interact with other people and animals—an alter of a human DID patient cannot perceive and interact directly with another alter of the same patient; there is nothing the second alter *looks like* from the point of view of the first; the first alter cannot reach out and touch the second. So how is it that I can reach out and touch other people and animals if they, like me, are analogous to alters of the universal will?

The key to making sense of this is rigor in interpreting the analogy: we are likening (a) a person with DID to (b) the universal will with something analogous to DID. But remember, unlike the case of the person, *there is no external world from the point of view of the universal will.* The latter is, *ex hypothesi*, all there is, all phenomena being internal to it. So we are comparing apples to bananas when we relate the person's life *in the outside world* to the entirely endogenous inner life of the universal will. It is much more apt to compare the latter with the person's *dream life,* for only then all experiential states *in both cases* are internally generated, without the influence of an outside world. This, and only this, is a fair analogy.

So what do we know about the dream life of a human DID patient? Can the patient's different alters share a dream, taking different co-conscious points of view within the dream, just like you and I share a world? Can they perceive and interact with one another within their shared dream, just as people can perceive and interact with one another within their shared environment? As it turns out, there is evidence that this is precisely what happens, as research has shown (Barrett 1994: 170-171). Here is an illustrative case from the literature:

The host personality, Sarah, remembered only that her dream from the previous night involved hearing a girl screaming for help. Alter Annie, age four, remembered a nightmare of being tied down naked and unable to cry out as a man began to cut her vagina. Ann, age nine, dreamed of watching this scene and screaming desperately for help (apparently the voice in the host's dream). Teenage Jo dreamed of coming upon this scene and clubbing the little girl's attacker over the head; in her dream he fell to the ground dead and she left. In the dreams of Ann and Annie, the teenager with the club appeared, struck the man to the ground but he arose and renewed his attack again. Four year old Sally dreamed of playing with her dolls happily and nothing else. Both Annie and Ann reported a little girl playing obliviously in the corner of the room in their dreams. Although there was no definite abuser-identified alter manifesting at this time, the presence at times of a hallucinated voice similar to Sarah's uncle suggested there might be yet another alter experiencing the dream from the attacker's vantage. (*Ibid*.: 171)

Taking this at face value, what it shows is that, while dreaming, a dissociated human mind can manifest multiple, concurrently conscious alters that experience each other from second- and third-person perspectives, just as you and I can shake hands with one another in ordinary waking life. The alters' experiences are also mutually consistent, in the sense that the alters all seem to perceive the same series of events, each alter from its own individual subjective perspective. The correspondences with the experiences of individual people sharing an outside world are self-evident and require no further commentary.

Clearly, our *empirical* grasp of extreme forms of dissociation shows that a DID-like process at a universal scale is, at least in principle, a viable explanation for how individual subjects arise within the universal will. Whether the cognitive mechanisms underlying dissociation are also *conceptually* understood today is but a secondary question: whatever these mechanisms may be, we know empirically that they *do exist* in nature and produce *precisely the right effects* to explain the illusion of individuality posited by Schopenhauer. In this regard—and in many others as well—Schopenhauer's metaphysics is empirically plausible.

The question now is: What counts as alters of the will? We know that human beings do. Animals do too, insofar as they are akin to us except in that they lack abstract representations. But what about plants, which react to stimuli but lack even perceptual representations? And what about inanimate objects, such as rocks and magnets, in which only the universal forces are manifest?

The first thing to notice is that Schopenhauer makes an uncompromising distinction between living organisms and inanimate objects:

> even between the smallest lichen, the lowest fungus, and everything inorganic there remains *a fundamental and essential difference*. (W2: 296, emphasis added)

He proceeds to explain that living organisms are defined by their *form*—which they maintain despite a constant exchange of material between their bodies and the environment, through feeding and secretion—whereas inanimate objects are defined precisely by the particular *material* that happens to constitute them at any one point in time.

Moreover, living organisms "are enclosed in a skin" (W2: 297)—i.e. in a boundary of some sort, an important point I shall discuss further in the next chapter—whereas an inanimate object is not. In the latter case, there isn't "anything [to] separate it from the outside world," and so inanimate objects "can easily be referred to fixed fundamental characteristics, which we call laws" (*Ibid.*). These natural laws or forces, in turn, are universal— *not* individuated—for

> a force of nature as, e.g., gravity or electricity, must manifest itself as such in *precisely the same way* in all its millions of phenomena … This *unity of inner being* in all its phenomena … is called a law of nature. (W1: 133, emphasis added)

Therefore, insofar as seemingly individual inanimate objects can be fully reduced to universal laws, they have unitary inner being and aren't *actually* individuated at all; they can't have conscious inner lives dissociated from the rest of the universe.

The carving out of the inanimate universe into separate 'things' has no metaphysical basis, according to Schopenhauer (W2: 299). These 'things' are just cognitively salient 'protuberances,' as it were, of the unitary objectification of "eternal and omnipresent" (W2: 301) volitional states in the form of natural forces. Schopenhauer insists,

> the will, objectifying itself in *inorganic* nature, no longer appears here in *individuals* who by themselves constitute a whole, but in natural forces and their action (W2: 336, emphasis added)

The existence of distinct inanimate objects is thus merely *nominal*, the output of a cognitive function performed by the intellect and already elucidated by Kant in his *Critique of Pure Reason*: the intellect clusters sense data into bundles—each bundle constituting a seeming object—according to certain *purposes*. A web of cause-and-effect relationships is then laid on these bundles so to link them together in a cognitively coherent manner. This way, individual inanimate objects are epistemic tools—creations of the observing individual subject—that serve some cognitive purpose, as opposed to distinct entities of the inanimate world as it is in itself.

Now, if the individuation of inanimate objects is merely epistemic, then—again—they cannot correspond to alters of the universal will. After all, dissociation has an *ontic* character: whereas I *can* arbitrarily stipulate that a handle is integral to the mug it's attached to, I cannot just decide that the chair I am sitting on is integral to me as an individual subject, for, try as I might, I cannot immediately—i.e. without mediation—*feel* what happens in the chair (I can only *represent* the chair through my sense organs). Analogously, whereas I *can* arbitrarily stipulate that a hood isn't integral to the coat it is attached to, I cannot just decide that a patch of my skin isn't integral to me as an

65

individual subject, for, whether I like it or not, I *do* immediately feel what happens in the patch. Metaphysically speaking, thus, there are no such things as individual inanimate objects that could correspond to dissociated alters of the universal will.

Admittedly, if read cursorily, some passages of Schopenhauer's may *seem* to suggest that individual inanimate objects have private conscious inner lives of their own; *but only if read cursorily.* Consider, for instance, this passage:

> Now let us consider attentively and observe the powerful, irresistible impulse with which masses of water rush downward, the persistence and determination with which the magnet always turns back to the North Pole, the keen desire with which iron flies to the magnet, the vehemence with which the poles of the electric current strive for reunion (W1: 118)

Commenting on this passage, Janaway contends that Schopenhauer cannot possibly mean "that iron really desires anything, or that water rushes because it wants to" (2002: 37). For Janaway, the notion that there is something it is like to be an individual river, or a magnet, or a piece of iron, separate from the rest of the inanimate universe, is "merely embarrassing" (2002: 36). And indeed, Schopenhauer clearly doesn't mean this at all, but not for the reason Janaway seems to assume: the problem is not that the inner essence of inanimate nature—even when taken as a whole—cannot be experiential, but that, ultimately, there are no separate inanimate objects such as rivers, magnets or pieces of iron.

With the passage quoted above, Schopenhauer is saying merely that the *universal* laws of nature—as locally *manifested* in the effects of gravitation in particular bodies of water and of electromagnetism in particular magnets and pieces of iron—are the objectification of equally *universal* volitional states. There isn't anything it is like to be a river or a magnet in and of itself; there is only something it is like to be *the inanimate universe as a whole*. These *universal* volitional states manifest themselves as gravitational and electromagnetic attraction, which, in turn, have local *effects* in what our *intellect* cognizes as particular bodies of water and magnets. If you read the passage above again, with this clarification in mind, you will see what Schopenhauer is trying to say.

The point here is important, for some construe Schopenhauer's metaphysics to imply some form of constitutive panpsychism (e.g. Koch 2014)—i.e. the notion that e.g. subatomic particles are already fundamentally conscious in and by themselves, and then somehow combine to form higher-level subjects of experience such as you and me. I find such an interpretation as perplexing as it is untenable, for Schopenhauer insists repeatedly that there is only *one* universal will or consciousness, the "*one* eye of the world which looks out from all knowing creatures" (W1: 198, original emphasis), "the one being" of nature (W2: 321). There are no fundamentally separate units of consciousness corresponding to individual inanimate objects such as subatomic particles; there aren't even individual inanimate objects outside mere representation of the intellect to begin with.

There is thus no way around our earlier conclusion: inanimate objects do *not* count as alters of the universal will, for there is no metaphysical basis for their individuation. Only living organisms have private—i.e. dissociated—conscious inner lives of their own.

But then, do *all* living organisms count as alters? The answer is yes, insofar as all living organisms (a) maintain their form while constantly exchanging material with the environment and (b) are enclosed in some sort of (porous) boundary that demarcates them from this environment, so that they (c) can have private experiential states dissociated from the environment.

According to Schopenhauer, even plants have dissociated experiential states, although very simple:

> We must picture to ourselves the *subjective* existence of the plant as a weak analogue, a mere shadow of comfortable and uncomfortable *feeling*; and even in this extremely weak degree, the plant knows only of itself, *not of anything outside it* (W2: 278, emphasis added).

This implies a dissociative boundary that carves out the inner conscious life of the plant from its outside environment.

According to Schopenhauer, the level of dissociation of a living organism from the universal background of uniform laws can be discerned in the strength of the organism's 'character': its degree of *differentiation* from said background. Whereas the inanimate universe complies with overarching, uniform causal patterns, a dissociated alter has its own local, idiosyncratic behavioral tendencies and dispositions—its own impulses or 'laws of behavior,' so to speak—which set it apart from the behavior of the rest of the universe (W1: 130-139). It is dissociation that enables such localized differentiation.

Mainstream present-day thought would have us believe that the character of living beings, too, is ultimately determined by the action—e.g. in brains—of the same universal forces that determine the behavior of inorganic nature. This, if true, would invalidate the notion of character as used by Schopenhauer, for the latter entails that a living organism's behavior is *not* reducible to universal physical laws:

> we shall certainly find in the organism traces of chemical and physical modes of operation, but we shall never explain the organism from these, because it is by no means a phenomenon brought about by the united operation of such forces, and

therefore by accident, but a higher Idea that has subdued these lower ones through *overwhelming assimilation*. (W1: 145, original emphasis)

In other words, the character of a living organism—as demarcated by dissociation—is not the bottom-up aggregate of the operation of universal physical forces, but results instead from the overwhelming top-down influence of higher-level volitional states on the behavior of the organism. Counter to current mainstream thinking as it may be, this postulate of Schopenhauer's is not—and arguably can *never* be—disproven.

Indeed, today it is simply *assumed* that the basic laws of physics—demonstrated to hold at a microscopic level—are solely responsible for the behavior of macroscopic systems in a causally-closed manner. In other words, it is merely assumed that there is nothing about the weather, astronomic phenomena, my or your behavior that cannot be explained in terms of the properties of subatomic particles. But this assumption cannot be verified: due to an exponential explosion of mathematical dimensionality and complexity, a subatomic-level simulation of, say, my or your nervous system—which could be compared to our observed behavior in real life—would take, to put it very mildly, impractically long to perform. Even if we assumed barely realistic future advancements in computing power, the complexity of the task is just mindboggling. So we do not really know if observations at a macroscopic level would turn out consistent with a microscopic 'Theory of Everything.' As acknowledged by Mile Gu *et al.*,

The question of whether some macroscopic laws may be fundamental statements about nature or may be deduced from some 'theory of everything' remains a topic of debate among scientists. (2009: 835)

More specifically considering the character of organisms, Robert Laughlin and David Pines (2000) argued that it is—and shall remain—impossible to demonstrate that the behavior of living beings can be fully modeled by a physical Theory of Everything and its associated equations:

> predicting ... the behavior of the human brain from these equations is patently absurd ... We have succeeded in reducing all of ordinary physical behavior to a simple, correct Theory of Everything only to discover that it has revealed exactly nothing about many things of great importance. (*Ibid.*: 28)

For all we *can* know, thus, Schopenhauer may be correct that character cannot be reduced to universal physical laws; it may indeed be a local differentiation from the uniform, universal background of forces.

A particular character can be discerned in a species as a whole (such as in the case of plants) or even in particular individuals (such as in the case of higher animals and human beings). In either case, however, there is always a degree of differentiation from the universal background of laws and forces, which is part of what defines organisms as individual subjects (W1: 130-139). Inanimate nature, on the other hand, embodies merely the universality of these laws and forces, uniformly applicable throughout spacetime (W1: 112-119).

Character is a manifestation of a local dissociative process unfolding in the universal will. Wherever character is discernible, there is an alter of the will. And because only living organisms are alters, the notions of dissociation, life and character are highly intertwined in Schopenhauer's metaphysics.

To sum it all up, *all living organisms are dissociated alters of the will*, each manifesting a character. But *only* living organisms—

*not* inanimate objects—are individual subjects. Some of these individual subjects are capable only of reactions to stimuli (plants); others are also capable of perceptual representations (animals); and some are capable even of abstract representations (humans). Nonetheless, all are alters of the universal will.

## Chapter 11

# An overarching conceptual framework

*But what does man possess that God does not have? Because of his littleness, puniness, and defencelessness against the Almighty, he possesses ... a somewhat keener consciousness based on self-reflection: he must, in order to survive, always be mindful of his impotence. God has no need of this circumspection, for nowhere does he come up against an insuperable obstacle that would force him to hesitate and hence make him reflect on himself.*
Carl Jung, in *Answer to Job* (1952)

Figure 3 summarizes the preceding discussions. The only ontological primitive in Schopenhauer's metaphysics is the (universal) will, depicted in the figure by an all-encompassing circle. Within the will, alters—depicted by smaller circles—form as a consequence of dissociative processes. Each alter has its own internal experiential states, encompassed by its respective dissociative boundary. The experiential states of the will that are *not* encompassed by any alter are referred to in the figure as the 'will-at-large.'

All alters are surrounded by the experiential states of the will-at-large, which—as we've seen in Chapter 9—exist in a (quantum) superposition outside spacetime. Also, as Schopenhauer insists, the notions of individuality and causality do not apply to the will-at-large. It consists instead of a unitary, indivisible (quantum-mechanically entangled) whole.

Perception results from the *interaction* between internal experiential states of the alter and external, superposed experiential states of the will-at-large (this, in fact, can explain quantum observation and the so-called 'collapse of the wave

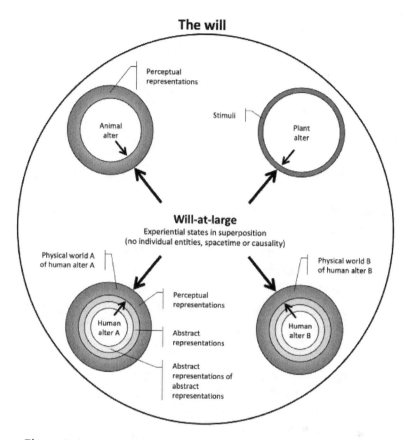

**Figure 3.** A conceptual framework for interpreting Schopenhauer's metaphysics.

function,' as I've elaborated upon in Kastrup 2017a). More specifically, an external experiential state *impinges* on an internal one—much like e.g. repressed, effectively dissociated emotions can surreptitiously impinge on our thoughts and produce new experiential gestalts as a result—thereby creating a new experiential state: the immediate feeling of sense impression. The latter, once processed by the intellect, becomes a perceptual representation. The mutually-pointing arrows at the dissociative boundary of each alter in Figure 3 depict this interaction between internal and external experiential states. The dark grey space between the arrows corresponds thus to perceptual representations. Clearly, the contents of perception are always *relative* to the individual subject.

Equivalent but simpler interactions between internal and external experiential states also give rise, in the case of plants, to stimulus apprehension. Indeed, Schopenhauer grants this equivalence between perception in animals and stimulus apprehension in plants:

> What susceptibility to light, in consequence of which it guides its growth in the direction of the light, is to the plant is the same in kind as knowledge [i.e. perceptual representation] to the animal, in fact even to man (W2: 284)

In the case of human alters, perceptual representations can be (further) re-represented in the form of abstract representations, which in turn can ground further levels of abstract representations. This corresponds to the higher levels of human meta-consciousness, depicted in Figure 3 by the concentric lighter-grey circles in human alters.

The white inner core of the alters, circumscribed by the respective layer(s) of representations or stimuli, depicts the alter's own *endogenous* experiential states—i.e. the endogenous feelings shown in Figure 1.

The physical world consists, according to Schopenhauer, of perceptual representations in the intellect of an observing individual subject. Therefore, as shown in Figure 3 and in compliance with Relational Quantum Mechanics (Rovelli 1996), each alter has its own physical world as a part of itself. Yet, these different physical worlds can be *similar to* and *consistent with* one another—just as multiple cars of the same make and model can be practically indistinguishable from one another—for all alters are immersed in the will-at-large and, as such, inhabit the same experiential environment. Solipsism is thereby avoided.

The physical world of an alter is an objectification of the experiential states of the will-at-large that 'surround' — in a cognitive, rather than a spatial, sense — the alter. But because, in addition to the inanimate universe, we also perceive other living organisms around us, some of these surrounding states must somehow (indirectly) correspond to other alters. This is possible because, as an alter $A$ interacts — through mutual impingement of experiential states across the respective dissociative boundary — with the will-at-large, some of the states of the latter accrue information about alter $A$. Then, when another alter $B$ interacts with the will-at-large, some of that accrued information about alter $A$ becomes reflected in experiential states constituting the physical world of alter $B$. Alter $A$ then becomes indirectly represented in the physical world perceived by alter $B$. In the language of representation — i.e. of objects in spacetime causally interacting with one another — we can say the exact same thing as follows: an organism $B$ perceives another organism $A$ because $A$ leaves an imprint in the world it shares with $B$, in the form of released photons, air vibrations, drifting scent molecules, etc.

What is the form of an alter's perceptual representation? As we've seen in Chapter 10, it is a living body, a notion that can be restated in a variety of different but equivalent ways:

(a) The internal experiential states of an alter are *presented* on the screen of perception of another — or even the same — alter as a living body.

(b) The internal experiential states of an alter are *objectified* in the physical world of another — or even the same — alter as a living organism.

(c) A living body is what the internal experiential states of an alter *look like* from across the alter's dissociative boundary.

Indeed, Schopenhauer explicitly links each alter to a living body:

> the subject of knowledge ... appears *as individual* only through his identity with the *body* (W1: 100, emphasis added)

Life is thus the physical appearance or objectification of dissociative processes in the will. It follows that death—i.e. the end of life—must be the physical appearance of the *end* of this dissociation—i.e. the reintegration of an alter into the will-at-large. And indeed, Schopenhauer says that

> death does away with the illusion that separates [an individual subject's] consciousness from that of the rest (W1: 282)

Isn't it clear that, for him, death represents the end of dissociation and life, therefore, a form of dissociation?

Schopenhauer's insistence that individual subjects correspond to organisms "enclosed in a skin" (W2: 297)—i.e. in a boundary of some sort—can also be construed as implying a dissociative process: as shown in Figure 3, the experiential states of an alter are carved out from the will-at-large and enclosed in a dissociative boundary. Since the objectification of an alter is its respective physical body, there must be something in the physical body that corresponds to the dissociative *boundary* of the alter. Put differently, there must be something the dissociative boundary *looks like* when represented as part of the physical body. In our case, this is indeed our *skin*, but also our retinas, eardrums, taste buds and mucous lining of the nose: *our sense organs are the objectification*—the perceptual representation—*of the dissociative interface through which our alter interacts with the will-at-large.*

The conceptual framework proposed in Figure 3 makes explicit that sense impressions—and, therefore, perceptual representations—only arise in the will as a consequence of dissociation: it is the formation of a dissociative boundary that allows for the non-associative impingement of experiential states beyond the boundary on experiential states within the boundary. This impingement—which constitutes sense impressions—is analogous to how e.g. a repressed emotion can impinge on a thought without having evoked the thought through an associative link. Our sense organs—the very vehicles of perception—are merely the extrinsic appearance of this dissociative boundary; they would not exist without dissociation and, thus, neither would perception.

Can we trace this crucial conclusion back to Schopenhauer's own words? We surely can:

> *knowledge* [i.e. perceptual representations] and plurality, or *individuation*, stand and fall together, for they condition each other. (W2: 275, emphasis added)

Now, since the brain is an integral part of the body, it, too, is the objectification or physical representation of certain experiential states of the alter: namely, those corresponding to what Schopenhauer calls the 'intellect' and 'reason.' This is why Schopenhauer speaks of both as functions of the brain (e.g. W2: 245-246): brain anatomy and physiology are what the intellect and reason *look like* from across their dissociative boundary. The rest of the body is what the less meta-conscious—or *not* meta-conscious—experiential states of the alter look like from across the boundary. Schopenhauer is clear in this regard in many passages, such as this:

> that which in self-consciousness, and hence subjectively, is the intellect, presents itself in the consciousness of other things, and hence objectively, as the brain; and that which in self-consciousness, and hence subjectively, is the will, presents itself in the consciousness of other things, and hence

objectively, as the entire organism. (W2: 245)

Janaway construes this correspondence between the intellect and the physical brain as Schopenhauer's "espousing a particularly blunt form of materialism" (2002: 48). This is rather baffling, as not only does Schopenhauer's argument entail or imply nothing of the kind—much to the contrary, as it should be obvious by now—but Schopenhauer also downright *ridicules* materialism, more than once, considering it e.g. reason for "a sudden fit of the inextinguishable laughter of the Olympians" (W1: 27). Indeed, insofar as materialism entails that experiential states are constituted or generated by particular arrangements of something outside and independent of consciousness, Schopenhauer is categorically *not* a materialist. The correspondence between intellect and physical brain that he talks about is one between essence and appearance, *both of which are experiential.*

The conceptual framework proposed in Figure 3 brings out the relationship between individual subjects and their surrounding environment left implicit in Schopenhauer's metaphysics. It also clarifies how the process of perception, in and of itself, arises and unfolds within the unitary will.

Most importantly, the proposed framework anticipates modern attempts to circumvent both the 'hard problem of consciousness' (Chalmers 2003) that afflicts physicalism and the 'subject combination problem' (Chalmers 2016) that afflicts constitutive panpsychism: by positing that universal consciousness—the will—is nature's sole ontological primitive, and that it is fundamentally unitary, the framework avoids both (a) the need to reduce consciousness to some other ontological category, and (b) the need to explain how fundamentally separate subjective points-of-view combine.

While circumventing these arguably insoluble problems, the framework still accommodates and makes sense of canonical empirical facts that would otherwise be at least difficult to reduce to universal consciousness. These facts are:

(a) we all seem to be separate beings inhabiting the same world;

(b) we can't change this world simply by wishing it to be different; and

(c) our inner experiential states correlate with measurable brain activity.

The framework proposed in Figure 3 reduces all these facts *without postulating any ontological category different from consciousness*:

(a) we are all dissociated alters immersed in the will-at-large, which explains why we seem to be individuals who inhabit the same world;

(b) we can't change the will-at-large merely by wishing it to be different because we are dissociated from it; and

(c) measurable brain activity correlates with our inner experiential states because the former is what the latter *look like* from across our dissociative boundary.

By explicating these facts coherently—a much more detailed and rigorous elaboration of which I have provided elsewhere (Kastrup 2018a)—the framework allows us to reduce the whole of empirical reality to the unitary will, just as Schopenhauer intended.

I acknowledge that this conceptual framework—unlike most of what has been discussed in previous chapters—is not explicit in Schopenhauer's argument. As such, it is both an interpretation and a further elaboration of Schopenhauer's thoughts. Nonetheless, I submit that it compellingly unlocks and reveals the overarching sense and coherence of Schopenhauer's metaphysical ideas. In other words, it brings Schopenhauer's various claims together in an internally consistent and satisfying manner that—short of a rather fanciful fluke—cannot but reflect Schopenhauer's underlying, implicit thinking. When one reads *The World as Will and Representation* under the light of this framework, it becomes difficult to imagine that Schopenhauer could have had anything substantially different in mind.

# Chapter 12

# Tackling objections to Schopenhauer's metaphysics

*I maintain that God is the immanent cause, as the phrase is, of all things, and not the transitive cause. All things, I say, are in God and move in God. ... However, as to the view of certain people that [my book] the Tractatus Theologico-Politicus rests on the identification of God with Nature (by the latter of which they understand a kind of mass or corporeal matter) they are quite mistaken.*
Baruch Spinoza, attempting to dispel—in a letter to Henry Oldenburg, written near the end of 1675—a misinterpretation of his metaphysics that, bafflingly, persists to this day

Having schematically laid out the overall structure of Schopenhauer's metaphysics, we are now well equipped to clarify what I believe to be some prominent misinterpretations of it.

Janaway argues that, because Schopenhauer characterizes the will as blind striving devoid of *deliberate* purpose, it unfolds "at a level beneath that of conscious thought" (2002: 35). He also claims that Schopenhauer

clearly does not think that organisms entertain any *conscious* purposes—for the will works 'blindly' (2002: 45)

The confusion here—a common one, as we've seen in Chapter 5—is in mistaking meta-consciousness for consciousness proper. This is forgivable insofar as even Schopenhauer conflates the terms. Nonetheless, allow me to reiterate: whereas meta-consciousness is indeed required for *deliberate* choices—the sort we experience when we plan a trip or choose a mortgage package—conscious volition can be experienced in an immediate manner, without re-representation, such as when we choose our left or right foot to take our first step in the morning.

A very illustrative example is our will to breathe, for it can be experienced both meta-consciously and merely consciously.[20] For instance, up until this moment you have been consciously experiencing your will to breathe, but without being aware *that* you want to breathe. The will to breathe is ordinarily autonomous— Schopenhauer would call it "immediate, necessary, and certain" (W1: 12)—unfolding without re-representation. However, the moment your attention is redirected appropriately, you re-represent it: you realize *that* you want to breathe. At this point, you can even deliberately hold your breath. So whether volition takes place with or without re-representation, it can be conscious—i.e. entail qualia, 'what-it-is-likeness'—in either case.

In its raw, primordial configuration, the will is clearly not meta-conscious. As such, it indeed entails blind striving; *but only insofar as our own will to breathe is also, ordinarily, a form of blind striving*: it unfolds the way it does as a consequence of how we function, not as a result of deliberation. Just as our will to breathe isn't actually unconscious, so the raw will isn't unconscious either. The blind striving described by Schopenhauer excludes meta-consciousness, not consciousness proper.

Although our will to breathe can be experienced both meta-consciously and merely consciously, the volition driving most of our autonomous biological functions—think of the beating of the heart, the activity of the glands and bodily metabolism in general—unfolds virtually always below the reach of introspection: except perhaps for some elusive, prodigious meditator somewhere in the Himalayas, the rest of us mortals can't experience it meta-consciously. *Yet it too*—according to my interpretation of Schopenhauer's metaphysics—*is conscious*.

Janaway makes a meal of Schopenhauer's warning (W1: 110-111) that we must extend our understanding of the will beyond ordinary human volition, so to grasp its manifestation in nature at large (Janaway 2002: 36-37). Bewilderingly, he implies that Schopenhauer doesn't clarify what this extension entails and, therefore, the door is supposedly left open to other interpretations of the will, besides conscious volition. Yet, Schopenhauer in fact *immediately* and *comprehensively* clarifies what he means: in humans, the will is

> guided by *knowledge*, strictly according to *motives*, indeed only to *abstract motives*, thus manifesting itself under the guidance of *reason*. (W1: 111, emphasis added)

In other words, human beings ordinarily experience the will as deliberate, introspectively accessible, *meta*-conscious volitional states. However,

> the will is also active where it is not guided by any *knowledge* [for] representation as *motive* is not a necessary and essential condition of the will's activity (W1: 114, emphasis added)

So the will in nature at large does not need to be accompanied by self-reflection—i.e. 'knowledge' or 'motive,' in Schopenhauer's terminology—as it ordinarily is in our case. Outside the field of self-reflection of human beings and higher animals, it still consists of conscious volitional states, but no *reflections* or re-representations thereof. *This is the required extension* of our understanding of the will beyond how we ordinarily experience our own volition. And because of it,

> if I say that the force which attracts a stone to the earth is ... will, then no one will attach to this proposition the absurd meaning that the stone moves itself according to a *known motive*, [merely] because it is thus that the will appears in man. (W1: 105, emphasis added)

How could Schopenhauer be clearer? To understand the will-at-large we must abstract self-reflection away from our conscious volition. In the inorganic world, the will is to be regarded as blind—though still *consciously felt*—impulses or instincts. That's it.

Schopenhauer is thus merely warning us not to limit our conception of the will to the higher levels of *meta-consciousness* at which we experience our own desires and fears. In the scheme of Figure 1, the point is that, in human consciousness, our (conceptual) knowledge of feelings eclipses our immediate experience of endogenous feelings; but in "weaker, less distinct phenomena" (W1: 111) of nature, the latter either dominates or constitutes the sole mode of acquaintance with the will. That Janaway seems to have missed Schopenhauer's extensive elaboration of this point (W1: 110-119) is odd.

Regarding Schopenhauer's account of the process of perception and its mediation by the bodily senses, Janaway offers the following objection:

> [According to Schopenhauer,] we apprehend some bodily state. The intellect then ... projects as cause of the sensation a material object 'outside' in space ... but [this] is troubling. For one thing, where do bodily sensations come from? *They must surely be originally caused in the body* by something *prior to the operation of the intellect*, but Schopenhauer does not discuss what the prior *cause* might be. (Janaway 2002: 21-22, emphasis added)

Of course he doesn't, for there cannot be any such *cause* prior to the intellect. We must ponder the question from within Schopenhauer's own logic and terms: the material body is just representation and so is causation. When he speaks of perception as being caused by a change in our bodily state, he is merely describing the process of perception *from the perspective of its representation*. In other words, he is describing how the process-in-itself, so to speak, *presents* itself in our intellect. But the process-in-itself is not material; it is not representation; it entails no physical body. Instead, it consists of the mutual impingement of experiential states across a dissociative boundary, as described in the previous chapter with reference to Figure 3.

Contrary to what Janaway seems to suggest in the passage quoted above, it makes no sense to look for a causal explanation *for the process-in-itself*, for we cannot reduce the process-in-itself to its representation. Attempting to do so inverts the logic of Schopenhauer's metaphysics. To understand perception as it is in itself we must think in terms not of physical causation, but of the experiential dynamics of the will, as illustrated in Figure 3. These dynamics are then *presented* to us in the form of physical things—e.g. photons, air pressure oscillations, drifting scent molecules, etc.—interacting with our skin and other sense organs in spacetime, which we then translate into a causal explanation by means of abstract reasoning. The representation of the process of perception in terms of particles, forces and causal action exists only in our respective physical world—see Figure 3 again—as an *image*, an *appearance*, of the actual, non-physical process of mutual impingement of experiential states.

If we continue to think in terms of the dynamics of the will itself—instead of its appearance on the screen of perception—Janaway's next question should strike us as trivial:

> Schopenhauer ... never claims that the will as thing in itself is a cause. But then what is the relationship between the world in itself and the things and events that lie within our empirical knowledge? (Janaway 2002: 39)

It is that illustrated in Figure 3: the dynamics of experiential states across the two sides of a dissociative boundary create, at the point where they interact with or impinge on one another, the perceived things and events that lie within our empirical knowledge—i.e. our respective *physical world*. Schopenhauer doesn't consider this experiential impingement 'causation' because he reserves the term for the behavioral nexus of the physical world, which is the *representation* of the impingement. Each alter has its own physical world as part of itself, within which we can speak of causation. But beyond the alters there are just the superposed experiential states of the will-at-large, not a physical world—i.e. not causation.

Janaway argues that Schopenhauer's claim that experiential states are the inner essence of the world—even the inanimate part of the world—is likely to be "dismissed as fanciful" (2002: 35), "something ridiculous" and even "merely embarrassing" (*Ibid.*: 36). He suggests that to salvage Schopenhauer's metaphysics we have to interpret the term 'will' as a rhetorical metaphor (*Ibid.*). Whilst acknowledging that the will must still be understood in terms of our privileged access to ourselves—i.e. in terms of what it is like to be us—he claims that

> we must enlarge its sense at least far enough to avoid the barbarity of thinking that every process in the world has a mind, a consciousness, or a purpose behind it. (Janaway 2002: 37)

This is a subjective and rather arbitrary value judgment, as Janaway doesn't argue for or substantiate it. Instead, he seems to be taking the mainstream physicalist commonsense—according to which consciousness cannot be anything other than an epiphenomenon of material brain function—for granted. If so, he is simply begging the question: since the point in contention is precisely one of metaphysics, it is circular reasoning to take a non-trivial metaphysical conclusion for an axiom.

Setting aside vulgar beliefs and biases, there is absolutely nothing fanciful, ridiculous, embarrassing or barbaric about the notion that the inner essence of everything consists of experiential states. This is a coherent postulate integral to a number of ontologies proposed and debated in analytic philosophy today, such as the many variations of panpsychism, cosmopsychism and idealism.

The final objection raised by Janaway reflects the subtlety of an important point. According to Schopenhauer, underlying and grounding each mortal individual subject is the "pure subject of knowing" (W2: 371). Unitary, universal and immortal, this pure subject of knowing is what "remains over as the eternal world-eye" (Ibid.) after individuality is abolished. Janaway then remarks that

Schopenhauer's attitude to this pure subject of representation is ambivalent. On the one hand, he says "Everyone finds himself as this subject" ... At the same time, however, each of us is an individual distinct from others. (2002: 51)

He has serious grievances about this ambivalence:

The problem, bluntly, is this: is my 'real self', or 'the kernel of my inner nature', something that attaches to the finite individual that I am, or is it the thing in itself, beyond space, time, and individuation altogether? ... *Schopenhauer seems to stumble into a quite elementary difficulty* [here]. (Ibid.: 68, emphasis added)

The difficulty, if any, seems to be Janaway's with figuring out what Schopenhauer is saying. Fortunately, present-day analytic philosophy — with its knack for eliminating even the most benign and easily reconcilable ambiguities—comes to his rescue.

In his attempt to circumvent the subject combination problem of panpsychism (Chalmers 2016), Shani posits that a unitary universal consciousness allocates its myriad experiential states to multiple "relative subjects" (2015). His hypothesis is, in a general sense, analogous to the idea of the will undergoing dissociation into multiple alters. Shani then remarks that all relative subjects inherit the "core-subjectivity" of universal consciousness by virtue of being segments of the latter (2015: 426), just as alters remain segments of the will. This core-subjectivity is the "dative … of experience [i.e.] that to whom things are given, or disclosed, from a perspective" (*Ibid.*, emphasis added). In other words, core-subjectivity is what remains of a relative subject when its experiential *contents* cease; it is pure subjectivity or 'Iness,' identical across all relative subjects.

We can directly transpose this to Schopenhauer's metaphysics: we are all the eternal will *in the sense that we, as alters, necessarily inherit the core-subjectivity of the will*. The dative or recipient of experience underlying each and every individual subject is identical, and identical to that of the will as a whole—the sole fundamental subject—which is thus "whole and undivided in every representing being" (W1: 5). Schopenhauer describes precisely this when he refers to the pure subject of knowledge as "that *one* eye of the world which looks out from *all* knowing creatures" (W1: 198, emphasis added), the "eternal world-eye" (W2: 371). The pure subject of knowing is *the subjectivity behind the 'eye' itself, not what the eye happens to see* from the individual perspective of any particular alter. If you and I were to become completely amnesic while in an ideal sensory deprivation chamber, for at least a moment all that would be left in both our inner lives would be this core-subjectivity, this 'Iness,' identical in both you and me.

At the same time, the fact that we are each a different alter confers upon each of us a unique, localized and restricted perspective within the activity of the will—an idiosyncratic point of view or window into the experiential environment surrounding us—which provides each of us with a particular set of contents of perception. Moreover, as alters, we each also have particular, private, endogenous volitional states concomitant with our instinctive desire to survive. In this sense, we are each a differentiated individual subject—mortal by virtue of the inexorable end of our respective dissociative process—constituted by private, idiosyncratic experiential states dissociated from the rest.

Schopenhauer is quite explicit about this double identity, stating, "we can attribute to everyone a twofold existence" (W2: 371). Consequently—and entirely reasonably—he also uses the word 'I' in two difference senses, depending on context. Nonetheless, Janaway still has a problem with it:

> Schopenhauer has previously told us that 'I' refers to the material, striving, human being ... which would not exist were it not for ... his or her bodily organs. But how could anything to which 'I' refers remain if the human being ceased to exist, taking with it the subject's consciousness? (2002: 108)

Given the preceding clarifications, it should be fairly easy to see that the pronoun 'I' is, in fact, entirely appropriate for referring to both of the identities or modes of existence of a human being: when referring to the pure subject of knowledge, 'I' denotes the *contentless* recipient of experience, immortal and identical across all individual subjects, which "remains unimpaired when [the individual subject] becomes extinct in death" (W2: 239). Indeed, in his mode of existence as the pure subject of knowing, "man is nature herself" (W1: 276).

On the other hand, when referring to a human being's mode of existence as an alter of the will, the pronoun 'I' denotes a particular set of experiential *contents* given to the universal recipient: the individual desire to survive in its many manifestations, as well as a conceptual mental narrative one identifies oneself with, both of which are constituted by experiential states circumscribed by the alter's respective dissociative boundary. This latter 'I' disappears upon death, in a way analogous to how a dream avatar vanishes when we wake up, or to how the alters of a human DID patient become re-integrated into the host personality upon a cure.

Schopenhauer's ambivalence is thus far from an expression of "confusion"—as Janaway rather daringly claims (2002: 68), given the stature of the philosopher he is referring to—but a sophisticated and nuanced effort to cater to people's various intuitions regarding what is meant by 'I.'

# Chapter 13

# Platonic Ideas and excitations of the will

*[I]f that species of motion which we term vibrations, can be shown, by probable arguments, to attend upon all sensations, ideas, and [bodily] motions, and to be proportional to them, then we are at liberty to make vibrations the exponent [i.e. ground] of sensations, ideas, and motions ... [The] power of forming ideas, and their corresponding miniature vibrations, does equally presuppose the power of association. For since all sensations and vibrations are infinitely divisible, in respect of time and place, they could not leave any traces or images themselves, i.e. any ideas, or miniature vibrations, unless their infinitesimal parts did cohere together through joint impression; i.e. association.*

David Hartley, in *Observations on Man, his Frame, his Duty, and his Expectations* (1749)

We have so far talked about both the *will* and its experiential states. We have also seen that the will is not a thing in the physical world, but irreducible subjectivity. How exactly, then, are we to think of its *states*? Schopenhauer's answer is cogent:

By virtue of the simplicity belonging to the will as the thing-in-itself ... its *essential nature* admits of no degrees, but is always entirely itself. Only its *stimulation or excitement* has degrees, from the feeblest inclination up to passion (W2: 206, original emphasis)

Therefore, the experiential states of the will are the outcome of its (self-)*stimulation* or *excitement*. We can visualize the will as an excitable substrate and an experiential state as a localized pattern of *vibration*—a 'note,' so to speak—of such substrate.

This, of course, is a metaphor, as the will is outside spacetime and cannot, as such, consist of a vibrating substrate. But it does provide a handy descriptive tool that Schopenhauer himself doesn't shy away from using. For instance, he characterizes the will as the "omnipresent substratum of the whole of nature" (W2: 326), claiming that "music is as immediate an objectification and copy of the whole will as the world itself is" (W1: 257). But music is the play of notes, the unfolding of patterns of vibration. For Schopenhauer, the ebb and flow of the will—the dance of existence itself—has the vibratory nature of a symphony. The will—the vibrating substrate of all existence—is the sole instrument playing this symphony.

Construing experiential states to correspond to excitations of the will is powerful: it allows us to make sense of the wide qualitative differences between, say, love and fear, suffering and bliss, red and blue, without requiring anything other than the will itself. Different patterns of excitation—'notes'—are what accounts for the different qualities of experience, even though the 'instrument' that plays these notes is always one and the same, "always entirely itself."

*For the same reason that there is nothing to ripples but the water in which they ripple, there is nothing to the myriad different experiential states of the will but the will itself,* sole member of Schopenhauer's reduction base. It is the unfathomable variety of the will's *behaviors*—in the form of its self-excitations—that leads to the complexity of nature.

Against this, one could use something analogous to David Hume's argument for the non-existence of mind. Will Durant framed the argument as follows:

We know the mind, said Hume, only as we know matter: by perception, though it be in this case internal. Never do we perceive any such entity as the "mind"; we perceive merely certain ideas, memories, feelings, etc. The mind is not a substance, an organ that has ideas; it is only an abstract name for the series of ideas; the perceptions, memories and feelings *are* the mind; there is no observable "soul" behind the process of thought. The result appeared to be that Hume had as effectually destroyed mind as Berkeley had destroyed matter. (2006: 334-335)

Transposing this to Schopenhauer's metaphysics, the argument could go as follows: we know the will only through our affections—i.e. particular experiential states. Never do we become *directly* acquainted with the entity that supposedly grounds these states, as water grounds ripples. Therefore, to infer the existence of that whose excitations are our affections—i.e. the non-stimulated will—is allegedly as precarious as to infer a physical world independent of representation.

To see why this argument fails, remember first that the mental image of a vibrating will is but a metaphor. The will is not a thing, or a physical substrate, or a "substance" in Durant's words, which can vibrate, but pure subjectivity instead. Assuming that the word 'excitation' is less committal to spatio-temporal extension, it is more accurate to speak of experiential states as *excitations of pure subjectivity.*

Now, if anything is certain in nature, it is that *a subject of experience*—whatever it may essentially be, and independently of any particular definition of self—*exists*; otherwise you wouldn't be there consciously reading these words. The existence of an experiencing subject is the starting point of our knowledge and nature's sole given; everything else is a theory (Kastrup 2018b). This subject is the recipient of experience—i.e. that to which qualities are disclosed from a perspective. All experiences, regardless of their particular qualities, entail the presence of this subject, of an experienc*er*. As Galen Strawson put it, not even a sensible Buddhist would reject such a claim (2006: 26).

Schopenhauer's point is thus simply that particular experiential states consist of particular patterns of excitation *of the subject itself*. What is given or disclosed to the recipient of experience are *its own excitations*. This way, nothing but nature's sole given—i.e. subjectivity—is required to explicate everything else.

Allow me to belabor this important point. That subjectivity—'what-it-is-likeness'—is inherent to *all* experiential states, regardless of their bewildering qualitative diversity, unites them in an ontological category referred to as 'phenomenality' in today's analytic philosophy. Therefore, when Schopenhauer posits that experiential states are but excitations of a universal subject—i.e. the will—*he is seeking to reduce the whole of nature to its sole given ontological category*. After all, for the same reason that there is nothing to a vibrating guitar string but the string itself, there is, in this case, nothing to an experiential state of the will but the will itself. Schopenhauer's metaphysics is thus as parsimonious and epistemically reliable as any metaphysics could possibly be, as far as its reduction base is concerned.

Contrast this to mainstream physicalism: there, the reduction base consists of theoretical abstractions that, *by definition*, are essentially different from subjectivity—i.e. fundamentally other than nature's sole given (Kastrup 2018b). After all, at a fundamental level, matter/energy isn't—according to mainstream physicalism—a subject of experience; only particular *arrangements* thereof allegedly are, epiphenomenally. Still at a fundamental level, matter/energy has only abstract, *quantitative* properties—such as mass, charge, momentum, spin and spacetime position—but no concrete, *qualitative* properties. The latter, of course, are all we can know directly.

Indeed, as Bishop Berkeley argued and Will Durant alluded to in the quote above, we only have access to the contents of perception—which are *experiential* in nature—not to matter/energy outside experience. Unlike subjectivity, a physical world fundamentally outside and independent of consciousness is but an explanatory model produced by our consciousness (*Ibid.*), which then fails to accommodate consciousness itself. This, as Schopenhauer wittily put it, "is the philosophy of the subject who forgets himself in his calculation" (W2: 313). When mainstream physicalism promises to eventually reduce consciousness to matter/energy, it is effectively promising to reduce consciousness to abstractions *of consciousness*.

Clearly, thus, the claim that inferring the existence of matter/energy—as entities outside consciousness—from perception is equivalent to inferring the existence of the unexcited will from experiential states is downright fallacious. In the latter case, no ontological category different from that *directly entailed* by knowledge is required, everything being reduced to a (universal) subject of experience; in the former case, one chases one's own tail by dreaming up a new ontological category and then attempting to reduce one's own subjectivity to this theoretical abstraction.

In conclusion, Hume's argument fails in connection with Schopenhauer's metaphysics.

The understanding that experiential states are excitations of the will is very useful for grasping how Schopenhauer integrates Plato's 'eternal Ideas'—the archetypes or primary templates of which, according to Plato, everything we perceive in the world is merely a distorted copy—in his metaphysics. Indeed, one of Schopenhauer's most nuanced metaphysical points is the relationship he draws between the eternal Ideas and the world-in-itself. Both

*always are but never become and never pass away. No plurality* belongs to them; for each by its nature is only one, since it is the archetype itself, of which all the particular, transitory things of the same kind and name are copies or shadows. (W1: 171, original emphasis)

According to both Plato and Kant, perceived things and phenomena, for being transitory, have no existence in themselves, but are merely the *expressions* of something eternal, universal and essential. Plurality and becoming exist only in the expressions, whereas what is essential simply *is* what it is, outside time and space. Plato calls this eternal essence the 'Ideas,' whereas Kant— whom Schopenhauer follows—calls it the 'thing-in-itself.'

That said, Schopenhauer does recognize a marked difference between these two concepts: whereas the eternal Ideas can be discerned through representation—Plato even talks of them as ideal *'Forms,'* which seems to immediately place them in the realm of perception—the thing-in-itself is precisely that which is *not* representation and has *no* form. So do the Ideas belong to representation or are they—as Schopenhauer claims—directly related to the thing-in-itself?

Admittedly, Schopenhauer sometimes uses seemingly contradictory language while trying to make sense of this. For instance:

> The Idea is only the immediate, and therefore adequate, *objectivity of the thing-in-itself*, which itself, however, is the will—the will in so far as it is *not yet objectified*. (W1: 174, emphasis added)

How can the Idea be the objectivity of something not yet objectified? Again, we must read Schopenhauer charitably, for even within the span of this single sentence he is using the notion of 'objectivity' in two related but different ways, which he distinguishes through the use of the qualifier 'immediate.' Indeed, Schopenhauer proceeds to explain that the eternal Ideas are the immediate objectivity of the will *in that they have only the primary property of objectification*: "that of being-object-for-a-subject" (W1: 175). However, unlike perceptual representations, the eternal Ideas have *none* of the other, secondary properties of objectification, such as spatio-temporal extension, plurality and causal relations.

As we've seen in Chapter 5, the property of "being-object-for-a-subject" involves self-reflection. Hence, for Schopenhauer *the eternal Ideas consist of a self-reflective apprehension of something eternal and essential about the will, without the cognitive mediation of spacetime and causality.* The Ideas "reveal to us most completely the essence of the will" (W1: 213), he says. This epiphany may be *facilitated* or *primed* by perceptual representation, in that Plato thought we could discern eternal Forms in objects of perception. However, perceptual representation itself is not part of the resulting insight. In other words, perceptual forms may *point* to the eternal Ideas, but the subsequent self-reflective apprehension of these Ideas is an act of cognition relating them *directly* to the subject of apprehension; the perceptual forms themselves do not survive in this final apprehension.

In the scheme of Figure 1, the eternal Ideas are archetypes of perceptual feelings that somehow percolate through to initial re-representation (step 3 of intellectual processing), without getting caught in the cognitive spatio-temporal scaffolding or the web of cause-and-effect relationships (steps 1 and 2, respectively).

Because this is all rather abstract, it is admittedly difficult to concretely visualize—and therefore *grasp*—what Schopenhauer is trying to *add* to our understanding of the thing-in-itself by relating it to the eternal Ideas. But here is where the excitation or vibration metaphor comes to our aid. Although Schopenhauer never utilized the metaphor in the way I am about to do now, I believe the interpretation below fits naturally into Schopenhauer's overall reasoning.

As we've just seen, a Platonic Idea is, for Schopenhauer, a direct, lightly self-reflective apprehension of something eternal and universal about the will, some *essential property* of it. This property is not itself perceptual representation, for the latter isn't essential, universal or eternal. Therefore, it can't be any particular pattern of excitation of the will, as represented on the screen of perception. Yet, it can somehow be *pointed to*, or *hinted at*, or *suggested by*, these patterns of excitation; there is something discernible in perceptual representation that in some sense *implies* essential properties of the will.

To solve this riddle in a satisfying, concrete way, we need—unavoidably, as no essential is amenable to direct visualization—to think metaphorically. We've seen that perceptual representations are patterns of excitation of the *individual* will—i.e. experiential states within an alter—that represent other patterns of excitation of the *will-at-large* (see Figure 3 again). We also know from basic physics that, when excited, a substrate tends to vibrate in one of its natural frequencies—i.e. according to one of its *natural modes of excitation*. For instance, when excited by plucking, a guitar string will tend to vibrate in certain notes, but not in others. Although arbitrary patterns of vibration can be actively induced in the guitar string through an external force, if excited and then left alone the string will vibrate in one of its natural frequencies, as determined by e.g. its elasticity and length. Indeed, unless forced to do otherwise by external, active interference, everything that can be excited will vibrate according to one of its natural modes.

*A substrate's natural modes of excitation are determined by the substrate's essential properties.* In the case of the guitar string, the essential properties are e.g. the string's elasticity and length. These properties can even be *deduced* from the note in which the string naturally tends to vibrate.

Now transpose this to the will-at-large: its essential properties determine its natural modes of excitation. The superposed experiential states of the will-at-large consist of its patterns of vibration and so would, *absent any interference*, correspond to these natural modes. But the mutual impingement between the experiential states of the will-at-large and those of the alters—as entailed by the process of perception, illustrated in Figure

3—*does* interfere with the vibrations of them all. What is then perceived by an alter is a *distorted representation* of the natural modes of excitation of the will-at-large.

*I submit that, in Schopenhauer's metaphysics, the eternal Ideas correspond to the natural modes of excitation of the will*—its basic templates of striving—*which, in turn, provide insights into the will's essential properties.* Perceptual representations, on the other hand, are but partial and distorted copies of these natural modes, resulting from interference patterns inherent to the process of perception. Nonetheless, they still insinuate, or point to, the eternal Ideas, merely for being partial and distorted copies thereof.

From the above, it follows trivially that—precisely as Schopenhauer claims—the Ideas are:

(a) Essential: they reveal what the will *inherently is,* for the intrinsic properties of the will are what determine its natural modes of excitation.

(b) Eternal: whereas particular patterns of excitation—like particular notes played on a guitar string—come and go, the will *always* tends to vibrate according to its natural modes, whatever patterns of vibration happen to be present at any one point in time.

(c) Unitary: natural modes of excitation are a *global* property of their substrate; they cannot be decomposed into constituent parts.

The vibration metaphor thus provides a concrete visualization of the nature of the eternal Ideas in Schopenhauer's metaphysics.

We've seen that the eternal Ideas correspond to the natural modes of excitation of the will-at-large. But before the first alter formed—i.e. before abiogenesis, the emergence of life from non-life—there was no distinction between the will and the will-at-large: they were one and the same. So the formation of alters—more accurately, of the dissociated *configurations* of the will that I've been calling alters—must correspond to natural modes of self-excitation of the will; otherwise, alters couldn't have formed in the first place: since the will is all there is, there was no external force to induce patterns of excitation in the will different from its own natural modes.

This explains why, for Schopenhauer, each living creature is a distorted copy of an eternal Idea underlying its species. My cats are just distorted copies of the eternal Idea of 'catness'; a particular dog is just a distorted copy of the eternal Idea of 'dogness'; etc. As such, it is

> of no importance whether we now have before us this animal or its progenitor of a thousand years ago … the Idea of the animal alone has true being (W1: 172)

So each particular kind of universal dissociation—whose objectification is a particular biological species—corresponds to a natural mode of excitation of the universal will.

Indeed, when Schopenhauer more specifically characterizes the will as the will-*to-live* (W1: 275), he is contending that *the formation of alters*—the rise of life—*is an intrinsic disposition of the will*. How and why this is so becomes clear when we interpret alter formation as an expression of the will's natural modes of excitation: the will 'wants' to create life for a reason analogous to why a guitar string 'wants' to play a certain note when excited.

The inanimate world—as perceptually represented—also partially and distortedly embodies *other* natural modes of excitation of the will-at-large. Therefore, according to Schopenhauer's metaphysics, the physical world as a whole—both organic and inorganic segments—is *symbolic*: it points to something beyond itself; it hints at a deeper reality beyond representations of any kind. Schopenhauer provides an example:

> the events of the world will have significance only in so far
> as they are the letters from which the Idea of man can be read
> (W1: 182)

So everything that unfolds in the physical world is a symbol
that points to an eternal Idea, just as letters and words denote
something beyond themselves. The physical world is akin to
a book to be read; it carries an implicit message. To cognize
this message we need *imagination*—i.e. *endogenous* experiential
states—so that we

> see in things not what nature has actually formed, but what
> she *endeavoured* to form, yet did not bring about, because of
> the conflict of her forms with one another ... imagination
> *extends the mental horizon* ... beyond the objects that actually
> present themselves (W1: 186-187, emphasis added)

For most ordinary people, however,

> the abstract concept of the thing is sufficient ... the ordinary
> man does not linger long over the mere perception, does
> not fix his eye on an object for long, but, in everything that
> presents itself to him, quickly looks merely for the concepts
> under which it is to be brought (W1: 187)

This sounds true enough: at least ordinarily, we do not bother to
see *past* the mere appearances. Instead of dwelling in an object
of perception to somehow discern the underlying template of
striving that gave it its form, we simply label it and move on
to something else. "What is that? Oh, it's *just* a bird," we say to
ourselves, and then shift our attention. Consequently, we become
limited to conceptualizations and rationalizations, failing to read

the letter for the sake of describing the envelope. We fail to attain insight into the ineffable essence—the underlying patterns of struggle of the will—the entire physical world is hinting at. This is unfortunate, for what the symbols of physicality are pointing to

is a world so rich in content that not even the profoundest investigation of which the human mind is capable could exhaust it (W1: 273).

So much for an ostensibly pessimist philosopher.

# Chapter 14

# The metaphysical meaning of life and suffering

*[I]f all nature presses towards man, it thereby intimates that man is necessary for the redemption of nature ... and that in him existence at last holds before itself a mirror in which life appears no longer senseless but in its metaphysical significance. ... Nature needs knowledge and it is terrified of the knowledge it has need of.*
Friedrich Nietzsche, discussing Schopenhauer's philosophy in his 1874 essay, *Schopenhauer as Educator*

Schopenhauer posits that the formation of alters—and the corresponding rise of representations—*serves* the will (W1: 275). This suggests that the latter has some intrinsic purpose or telos, which it strives towards through self-stimulation according to its natural modes of excitation. In fact, Schopenhauer is explicit about there being such a telos:

> Already striving towards its goal ... through its original laws themselves, the will works towards its final aim; and therefore everything that happens according to blind laws of nature must serve and be in keeping with this aim. (W2: 324)

But because the will, in its primordial or 'blind' configuration, is not self-reflective, this purpose cannot be premeditated or deliberate. The will strives towards its telos *without actually knowing*—in a meta-cognitive sense—*the motive for its struggles*. Instead, it strives 'blindly': "the final cause [i.e. goal] is a motive that acts without being known" (W2: 342). The will *feels* its goal but does not deliberate upon it; it doesn't even know *that* it has this goal.

We should thus think of nature's telos as an *instinctive disposition*, a "simple tendency of the original will" (W2: 328) to flow towards a goal or attractor, somewhat analogously to how rivers tend to flow towards the ocean. This instinctive disposition corresponds, as we've seen, to certain natural modes of self-excitation of the will: the will wants to reach its telos for reasons analogous to why a guitar string 'wants' to vibrate in its natural mode.

Schopenhauer himself uses the word 'instinct' for a disposition that isn't re-represented, providing several examples from the animal kingdom: spiders build their webs without meta-cognitive awareness of their goal; so do termites their hills, birds their nests, etc. (W2: 342). He goes on to compare the seemingly mechanical action of insects to that of a somnambulist (W2: 344), a remarkably prescient intuition: today we know that experiential states during sleep are characterized precisely by a reduction in meta-cognitive capacity (Windt & Metzinger 2007). In Schopenhauer's metaphysics, *instinctive action is thus action driven by a telos that is consciously felt but not meta-cognized.* This is what characterizes the actions of nature—except for re-representing organisms, such as you and I.

Schopenhauer goes on to suggest what nature's final cause or attractor may be:

> Through the addition of the world as representation, *developed for its service*, the will obtains *knowledge of its own willing* and what it wills (W1: 275, emphasis added)

So not only does the will have an instinctive telos, this telos has to do with the development of meta-cognitive self-awareness: the will instinctively wants to explicitly know what it wills and why it wills it.

Now, because only alters can develop meta-cognition[21]—

through their hierarchical layers of re-representation—this natural telos translates into the will-to-live: *life is the will's means to meta-cognize itself*, to raise its head above the otherwise overwhelming maelstrom of its own instinctive striving. Schopenhauer explains this in a marvelously aphoristic passage:

> what the will wills is always life, just because this is nothing but the presentation of that willing for the representation (*Ibid.*)[22]

The will's continual development of progressively more meta-conscious configurations, discussed in Chapter 7, is the embodiment of its teleological strife: the degrees of manifestation of the will correspond to steps or phases—each triggered and enabled by the previous, and each unfolding according to one of the will's natural modes of excitation—of its 'blind' attempt to achieve self-understanding.

Human beings are the apex of this progression, so our role is to bring into explicit, meta-cognitive awareness that which was previously known only instinctively. While misusing the words 'conscious' and 'consciousness' when he actually means 'meta-conscious' and 'meta-consciousness,' respectively, Erich Neumann echoes this view when he writes:

> Man's task in the world is to remember with his conscious mind what was knowledge before the advent of consciousness. (2014: 24)

To put it more directly, *the metaphysical meaning of human life is to achieve meta-conscious awareness of the dynamics of the will.* Through introspection we can meta-cognize it as it manifests within ourselves, and through contemplation of the eternal Ideas we can meta-cognize it as it manifests in the world beyond ourselves, both organic and inorganic realms. The metaphysical meaning of all *non*-human life, in turn, *is to enable the realization of the meaning of human life*. According to Schopenhauer, nature conspires—albeit in a confusing, instinctive, non-monotonic manner plagued by frequent missteps—towards developing meta-conscious humanity and then unveiling itself *to* humanity in the form of the eternal Ideas.

But the will's strife leads, in living creatures, to constant suffering. The restlessness that characterizes life is a reflection of the will's relentless pursuit of its goal. The more living creatures struggle, the more they suffer. Yet they can't stop struggling because—as they are nothing but local configurations of the will itself—they instinctively partake of the latter's desperate battle to figure itself out.

Schopenhauer opens the door to a solution, though. By attaining insight into *how* and *why* they suffer, living beings can take a substantial step towards easing their suffering, as any present-day psychotherapist would confirm. The problem is that Schopenhauer's prescription for what must happen next *seems*, at first, to contradict his entire metaphysics:

the last work of intelligence is to *abolish willing* … true salvation lies in the *denial of the will*. (W2: 610, emphasis added)

Before we reconcile this with Schopenhauer's claim that the will is all there is—and therefore can't really be denied—let us first discuss how the spirit of this prescription is related to a possible reduction of suffering.

The idea here is that *volition* is the source of all suffering. By craving certain things in the world and rejecting others—rejection is also a form volition, just in reverse—alters constantly set themselves up for suffering. If they fail to attain what they crave from the world, they suffer; if they succeed, they quickly get bored, begin to crave something else, and so suffer; if what they reject is imposed on them by the circumstances of the world around them, they suffer; if they manage to avoid what they reject, they fear that something *else* may be imposed on them by the world, against their will, and so they suffer; etc.

In what is essentially a Buddhist answer to this dilemma, Schopenhauer claims that only through the denial of their volition can alters stop suffering. And this denial is only achievable when alters *see through the suffering*, by means of meta-conscious introspection. By attaining the self-reflective insight that the root of all suffering is their desire, alters become able to put this desire in perspective and then overcome it.

This, however, may seem to pose a problem. Schopenhauer claims that we can become a "will-*less* subject of knowledge" (W1: 178, emphasis added). But how is that possible if the will is, ultimately, all there is? If the inner essence of the whole of nature is will, how can anything be will-*less*? How can the will deny itself out of existence?

To make sense of this, we need first to understand why Schopenhauer chose to call the experiential states at the foundation of all nature the 'will.' After all, why not just refer to them as, say, universal consciousness? Why qualify them further as states of a *volitional* nature?

The reason is two-fold: first, the experiential states of the will-at-large (see Figure 3 again) cannot entail—at least not purely— what is referred to in analytic philosophy as 'intentional content' (cf. e.g. Chalmers 1996: 19). In other words, they cannot be *about* a world out there, for the will-at-large is itself such a world; they cannot entail contents of perception, for perception belongs in alters. The will-at-large does not see mountains, or smell coffee, or hear symphonies; only alters do. So Schopenhauer had to characterize the primordial experiential states of the will in a manner that would exclude intentional content and leave room only for *endogenous* experiences unrelated to an external world. The primary example of these endogenous states is, of course, *volition*. For instance, if a human infant were to be isolated in an ideal sensory deprivation chamber from birth, it would arguably experience no intentional content but would still experience endogenous volitional states. The will, thus, is *will* because its primordial states—which precede alter formation—are of a purely endogenous character.

Second, it is an empirical fact that nature is dynamic: things *happen* in nature, events unfold; the sun shines, the wind blows, the rain falls, animals come and go. So whatever experiential states underlie nature, they must not only accommodate, but also *make sense* of, this dynamism; they must provide an *impetus* for action. And here is where volition comes in again: for nature to move, it must *want* something; otherwise, it would remain static. Without a volitional impetus, nothing would ever happen, for nature would be content to just remain in whatever state it happens to be. Schopenhauer clearly realized this, and so the will must, well, *will*.

Schopenhauer's characterization of what is effectively universal consciousness as volitional in nature seeks merely to accommodate the two points above. It shouldn't be seen as restrictive, for a variety of widely different endogenous—i.e. non-sensory—experiential states can be considered volitional in nature; perhaps even the entire gamut. Schopenhauer himself includes "terror, fear, hope, joy, desire, envy, grief, zeal, anger, [and] courage" (W2: 212). Anything that provides an impetus for action or change in the world, or even merely within an individual mind, can be looked upon as volitional.

Consequently, the denial of the will cannot be the denial of any particular *type* of endogenous experiential states—say, desire as opposed to anger—for all types can be looked upon as ultimately volitional. Yet it cannot be the denial of the whole will either, for the latter is all there is. Instead, what Schopenhauer means by the denial of the will is the overcoming of *endogenous feeling states of an individual alter* — thus *not* of the will-at-large — which arise in connection with the alter's survival instinct.

Indeed, because the will instinctively wants to understand itself, it stimulates itself into giving rise to dissociative configurations — alters, living creatures — that, over time, develop self-reflection. But there is nothing in Schopenhauer's argument to suggest that the primordial will, prior to alter formation, suffers. The original impetus is self-understanding, not the avoidance of suffering. Only living creatures suffer, as a side effect of their seeming separation or alienation from the rest of nature.

To overcome suffering, all that is thus needed is for individual subjects — alters — to overcome their own endogenous feeling states; for we have no reason to believe that the will-at-large suffers, and so its endogenous feeling states need not be supplanted. Now, what this overcoming entails is a nuanced point. Schopenhauer summarizes it as "a temporary preponderance of the intellect over the will" (W2: 367), but there is much more subtlety to it.

The alter's endogenous feeling states become supplanted *when it discerns the eternal Ideas* through the apparatus of perception (W1: 179), as discussed in the previous chapter. At this point,

> the person who is involved in this perception is no longer an individual ... he is *pure* will-less, painless, timeless *subject of knowledge*. (*Ibid.*)

There is a lot to unpack here. As we've seen earlier, the eternal Ideas correspond to a lightly self-reflective apprehension — through sense impressions — of universal templates of the will's objectification, but *without* the cognitive mediation of spacetime and causality. In other words, it involves step 3 of intellectual processing — i.e. initial re-representation, as illustrated in Figure

1—but bypasses steps 1 and 2. And as we've also seen earlier, the pure or timeless subject of knowledge corresponds to the *contentless* universal recipient of experience, the core-subjectivity of the universal will that underlies and grounds each alter. Here, thus, Schopenhauer is establishing a correspondence between the universal recipient of experience and the eternal Ideas: the latter can only be discerned when an alter dis-identifies with its particular endogenous feelings, there remaining only the former.

Notice how it all adds up: in order to become contentless, the alter has to subdue its endogenous volitional states; otherwise, it would still have salient endogenous contents. The way to achieve this is to observe an object in the world in such a manner as to *"lose ourselves entirely in this object"* (W1: 178, original emphasis). The "entire consciousness is [then] filled and occupied by a single image of perception" (W1: 179), which is the eternal Idea behind the object. And because eternal Ideas are universal and their apprehension independent of individual perspective— as opposed to individual objects, whose apprehension is defined by the particular spatio-temporal point of view of an alter within its physical world—once the alter's meta-consciousness becomes filled with them, the alter loses the ability to identify itself with any *non*-universal content of consciousness, thereby temporarily forfeiting its individuality.

Schopenhauer's recipe for subduing an individual's endogenous volitional states and reducing suffering is thus one of sensory overload:

the subject, by passing entirely into the perceived object, has also become the object itself, since the entire consciousness is nothing more than its most distinct image. (W1: 180)

It is because of this that

we are no longer able to separate the perceiver from the

perception, but the two have become one (W1: 179)

So by allowing an eternal Idea to completely fill the canvas of its meta-consciousness, an alter can supplant its own individual, endogenous volitional states—attendant upon its instinctive desire to survive as an organism—and thereby become the pure, painless subject of knowledge, who doesn't suffer. In the process of doing so, the alter also helps realize the metaphysical meaning of human life, insofar as contemplation of the eternal Ideas is a necessary step towards achieving the will's goal of explicit self-awareness.

According to the scheme of Figure 1, here is what happens: perceptual feelings encoding the basic archetypes of the will's objectification bypass steps 1 and 2 of intellectual processing, but are then subtly re-represented in step 3. This leads to a lightly self-reflective apprehension of these archetypes, which is the cognition of eternal Ideas. However, no further re-representation must take place, for

we do not let abstract thought, the concepts of reason, take possession of our consciousness, but, instead of all this, devote the whole power of mind to perception, sink ourselves completely therein (W1: 178)

The apprehension of the eternal Idea must be the *sole* and *complete* focus of attention of the alter, so as to supplant any intellectual processing of endogenous feelings (see Figure 1 again). Indeed, *the latter must not be re-represented at all*, so to free the entire field of self-reflection for the re-representation of the archetypes of the will's objectification. This is what the 'denial of the will' means in Schopenhauer's metaphysics, and it is entirely coherent.

Notice that, when Schopenhauer says that subject and object cannot be distinguished from one another during the apprehension of eternal Ideas, he is *not* implying that the subject-object split itself comes to an end—i.e. he does not mean that there is no re-representation. Indeed, he is very clear in asserting that an Idea "includes object and subject in like manner" (W1: 179) and that it has the fundamental property "of being-object-for-a-subject" (W1: 175). So initial re-representation *does* take place in the cognition of Ideas. However, because the meta-consciousness of the alter is filled *solely* with a reflection of the objectification archetype, it becomes qualitatively identical with the latter. To put it simply, subject and object still persist as distinct but indistinguishable entities, just as two cars of the same year, make and model can be indistinguishable while remaining distinct.

In summary, when Schopenhauer talks about the denial of the will he is referring merely to the subjugation of the *endogenous feeling states of an alter* by the alter's overwhelming apprehension of eternal Ideas through sense impressions. *But this apprehension itself also consists of (non-individual) states of the will.* As such, the 'denial of the will' isn't actually a denial of the will *as ground of being*, but merely a suppression of *particular experiential states* from the field of self-reflection of an alter. This is Schopenhauer's recipe for the temporary end—or at least alleviation—of human suffering, and for the achievement of life's metaphysical meaning.

# Chapter 15

# Concluding remarks

*Only that man's life is wasted who lived on so deceived by the joys of life, or by its sorrows, that he never became eternally and decisively conscious of himself as spirit, as self, or (what is the same thing) never became aware ... of the fact that there is a God, and that he, he himself, his self, exists before this God, which gain of infinity is never attained except through despair.*
Søren Kierkegaard, in *The Sickness unto Death* (1849)

We can now summarize Schopenhauer's metaphysics—at least as I interpret it—in fairly few words: all of reality consists, essentially, of one universal consciousness. Schopenhauer calls it the 'will' to (a) highlight the *endogenous* character of its original experiential states and (b) account for the *dynamism* of nature by attributing volitional impetus to these states.

In its primordial configuration, the will entailed no representations, as there were no individual subjects yet—i.e. no alters of the will. The experiential states of this primordial will did not include perception of a seemingly external world, for there was no such world yet. Instead, they entailed only endogenous *feelings*. Moreover, the *dispositions* or *impetus* inherent to these feelings triggered self-stimulation or self-excitation of the will, which in turn led to the latter's unfolding into the known universe.

The primordial impetus of the will is towards *self-understanding*: it yearns instinctively—i.e. *not* meta-cognitively—to figure out explicitly what it wills and why. Without self-understanding, the will drowns in the maelstrom of its own unfolding. It is this irresistible impulse that, through self-excitation according to the will's natural modes, led and still leads to the rise of living organisms: the images of local, dissociative configurations of the will—alters—that seemingly split off from the rest so to be able to contemplate it as object.

With life, there arose the world as *representation*—i.e. the image of the will as it presents itself from across a dissociative boundary—which, in turn, enabled *re*-representation. And because re-representation is the *sine qua non* of self-understanding, with it the will finally developed the potential for self-understanding that it instinctively seeks. The epitome of this development is human beings, capable of many layers of re-representation. Through us, and our meta-conscious contemplation of ourselves and of nature at large, the will learns about what it yearns and why.

The highest degree of the will's self-understanding is achieved through human contemplation of the *eternal Ideas*, which reveals the will's basic templates of striving or natural modes of excitation and, therefore, its essential properties. With focused attention and an initial re-representation that sets the world—as object—apart from the subject, human beings can apprehend these Ideas through perceptual feelings. This way, the world as representation comprises *symbols* of the eternal Ideas, pointers to something essential and immanent in all nature. Its purpose in the universal unfolding of the will is to be studied and deciphered for the attainment of the will's self-understanding.

Having apprehended the eternal Ideas, human beings can, *thereafter*, rationally process the corresponding insights. For this, we leverage the many layers of re-representation characteristic of our conceptual reasoning, so to replace the instinctive dispositions of the will with *deliberate purposes*. Through us, thus, the will attains a level of meta-cognitive self-control and a degree of freedom from the maelstrom of its own instinctive unfolding. Abstract representations overwhelm our endogenous feelings (see Figure 1), define our actions in their stead, and ultimately make sense of the world. This, for Schopenhauer, is the purpose of life.

It is unfortunate that such a cogent metaphysics — the foundation of the rest of the philosophical system for which Schopenhauer has become otherwise known — has been consistently misunderstood, misrepresented and dismissed for over two hundred years. Although Janaway's book was originally written in 1994, as late as 2018 — the year in which I started writing the present volume — the problem showed no sign of abating. For instance, in her popular biography of Friedrich Nietzsche, Sue Prideaux attempted to summarize Schopenhauer's metaphysics in this unfortunate passage:

> The representation is in a state of endless yearning and eternal becoming as it seeks unity with its will, its perfectible state. The representation may occasionally become one with the will but this only causes further discontent and further yearning. The human genius (a rare being) may achieve wholeness in the union of will and representation but for the rest of the human herd it is an impossible state in life, only to be achieved in death. (2018: 49)

It is remarkable in how many different ways this brief passage manages to be nonsensical. In just 77 words, Prideaux mistakenly claims that: (a) representations *yearn*, as opposed to being the image of yearning; (b) representations *have* will, as opposed to being objectifications of the will; (c) representations and will are not only separ*able*, but initially *separate*, as if there could be representations without will; (d) the will is a *state* of the representations, as opposed to their ground; (e) for most human beings representations and will cannot be *united*, as if most humans were philosophical zombies without inner essence... The density and severity of errors here is just overwhelming, even when the passage is read charitably. Whatever Prideaux thought she was describing, it has nothing to do with Schopenhauer's metaphysics.

Yes, Prideaux is a novelist and biographer, not a philosopher. But the moment one—*anyone*—sets out to summarize somebody else's work, one implicitly takes on the responsibility to either do it with a minimum of accuracy or *refrain* from doing it. Prideaux's seemingly innocent little passage about Schopenhauer will be read—probably has already been read—by many more people than will ever read Schopenhauer's own words. As a philosopher, I find this appalling. The idea that my own writings could one day receive this kind of treatment when I am no longer around to correct it is disturbing.

Persistent and insidious misunderstandings and misrepresentations, such as illustrated in the foregoing, have blinded us to an important fact: Schopenhauer's ideas in the sphere of ontology constitute a key nexus in the history of Western thought, linking our present-day efforts to circumvent insoluble conceptual problems—such as the hard problem of consciousness of physicalism (Chalmers 2003) and the subject combination problem of constitutive panpsychism (Chalmers 2016)—with those of our predecessors: Spinoza, Berkeley, Kant and Hegel.

For the two centuries since the release of *The World as Will and Representation*, our metaphysics has, sadly, continued to be in a state of disrepair. Worse yet, our intuitions have become so skewed that we—remarkably—have come to consider the untenable ontology of mainstream physicalism our least implausible option. Even the alternatives more openly discussed since the turn of the 21$^{st}$ century—such as the many variations of panpsychism—entail thought artifacts analogous to those that plague physicalism (Kastrup 2017b). Yet, a way to circumvent the insoluble problems of today's ontological menu has been available—if unrecognized—all along. Had Schopenhauer's metaphysics been properly appreciated earlier, our philosophical outlook today would perhaps be more mature.

This claim, of course, is contingent upon my interpretation of Schopenhauer's metaphysics being substantially correct. Critics, however, are bound to argue that such is not the case. Some may point out that there are passages of Schopenhauer's that seem to refute contentions made in the foregoing. To these critics I offer the following: whichever way one chooses to interpret key terms used by Schopenhauer, he cannot be read as though his use of these terms were fixed and strictly consequent. Regarded in this manner, his metaphysics will *always* be internally inconsistent, for his use of a term in one context will often contradict that in another.

Hence, as his very writing style demands, one must read Schopenhauer as if one were engaging in a colloquial conversation with him: his intended denotations must be charitably deduced from the overall context. This means that there will always be specific passages that, taken in isolation, will appear to refute one's interpretation. So either we decree that *no* interpretation of Schopenhauer's metaphysics is valid, or we have to find another way to assess the merits of an interpretation.

I submit that the way to do it is to gauge how well an interpretation brings Schopenhauer's various metaphysical contentions together in a coherent and mutually-reinforcing manner, while sticking to reasonable—albeit context-dependent—readings of all key terms. In other words, a good interpretation should explicate Schopenhauer's metaphysics in a way that makes *good overall sense*, without requiring that we artificially stretch the envelop of potential denotations of a term.

This, I contend, is precisely what the interpretation offered here achieves: it takes e.g. the word 'consciousness' ('*Bewusstsein*') to mean phenomenal consciousness or meta-consciousness, depending on the context, but never something beyond the word's reasonable envelop of denotations, such as, say, intelligence. It takes the word 'will' (*'Wille'*) to mean volitional experiential states—sometimes meta-cognitive and sometimes not, depending on the context—but never, say, force or some mysterious undefined entity or phenomenon only vaguely related to the word itself. And in reading terms in such a colloquially flexible but *sensible* manner, the interpretation offered here resolves all seeming contradictions in Schopenhauer's metaphysics. This ought to be a compelling sign that it is at least on the right track.

If so, this book carries with it the promise to rehabilitate and reintroduce Schopenhauer's metaphysics to a 21st-century audience exhausted by the insoluble problems of today's most popular ontologies. May it reopen more promising lines of reasoning and reacquaint us with vistas that, although perhaps counterintuitive to many of us today, made—for perfectly good reasons—much sense to our predecessors. May it help us realize that our (implicit) condescension towards our forerunners is misplaced and embarrassing: while unacquainted with much of today's science, philosophically they may have been ahead of us; indeed, they may have discerned truths we today dismiss out of mere prejudice, bad habit or even plain foolishness.

The way to the future of metaphysics may—perhaps surprisingly—be found in its past.

# Notes

1   Here I am deliberately avoiding the modern analytic terms 'phenomenal state' (Block 1995)—using the qualifier 'experiential' instead, even though I mean the same thing—to preempt confusion: Schopenhauer uses the term 'phenomenal' exclusively in connection with intentional content. By 'experiential states,' however, I mean to refer even to purely endogenous experiences devoid of intentional content. According to my usage here, a state is experiential if there is something "it is like to *be* in that state" (*Ibid.*: 227, emphasis added).

2   Schopenhauer repeatedly characterizes the world-in-itself as "*toto genere* different" from what we perceive.

3   The German word for 'representation' (*Vorstellung*) also translates as 'presentation.'

4   To preempt charges that my interpretation of Schopenhauer's words is founded on mistranslations, I shall occasionally also quote, in these notes, from the original German (Schopenhauer 1859). The first passage here in question states that the will is "*jenes Jedem unmittelbar Bekannte,*" which translates literally as "that immediately known to everyone." The second passage reads: "*Nur das Bewußtseyn ist unmittelbar gegeben,*" which translates literally as "Only consciousness is immediately given." The word '*Bewußtseyn*' ('*Bewusstsein*' in modern German) unambiguously means consciousness. And because the word '*unmittelbar*' (immediately, or without mediation) is the defining adverb in both passages, the unavoidable implication is that there is a sense in which the will *is* consciousness.

5   In Schopenhauer's original German (1859): "*was ich als anschauliche Vorstellung meinen Leib nenne, nenne ich, sofern*

*ich desselben auf eine ganz verschiedene ... Weise mir bewußt bin, meinen Willen ... der Leib noch in einer ganz andern, toto genere verschiedenen Art im Bewußtseyn vorkommt, die man durch das Wort Wille bezeichnet.*" This translates literally as: "What as graphic representation I call my body, I call my will insofar as I am conscious of it in a very different way. ... The body still appears in consciousness in a completely other, *toto genere* different manner, which one indicates by the word will." So, again, there is nothing wrong with the Payne translation quoted in the main text.

6   In Schopenhauer's original German (1859): "*das innere, einfache Bewußtsein.*" So here again we have the unambiguous German word '*Bewusstsein,*' consciousness.

7   "Besides the will and the representation, there is absolutely nothing known or conceivable for us." (W1: 105)

8   "the will ... considered as such and apart from its phenomenon ... lies outside time and space" (W1: 128).

9   "my body is the only object of which I know not merely the one side, that of the representation, but also the other, that is called will" (W1: 125).

10  Other translations seem to distort Nietzsche's initial clarification of what he means by 'consciousness.' Josefine Nauckhoff's, for instance, reads:

> The problem of consciousness (or rather, of *becoming conscious of something*) first confronts us ... (Nietzsche & Nauckhoff 2001: 211, emphasis added)

Even Walter Kaufmann translated it thus:

> The problem of consciousness (more precisely, of *becoming conscious of something*) confronts us only ... (Nietzsche & Kaufmann 1974: 297, emphasis added)

Nietzsche's original German, however, is hardly ambiguous:

*Das Problem des Bewusstseins (richtiger: des Sich-Bewusst-Werdens) tritt erst dann vor uns hin* ... (Nietzsche 1919: §354, emphasis added)

The highlighted segment means literally 'becoming conscious of oneself.' The verb *'werden'* (to become) is transitive—not obligatorily reflexive—so Nietzsche's use of the reflexive pronoun *'sich'* is deliberate: he is saying that what the subject becomes conscious of is itself. A possible explanation for Nauckhoff's and Kaufmann's translation is that they thought of more complete sentences that include an object or a subordinate object clause, such as *'er wird sich bewusst dass ...'* ('he becomes aware *that* ...'). But here (a) there is no object or subordinate object clause; and (b) even if there were, the corresponding sentence including the reflexive pronoun *'sich'* would still denote self-reflection—i.e. the ability to report the respective content of one's consciousness to oneself.

11  For the avoidance of doubt, here is Schopenhauer's original German (1859): "*Selbstbewußtseyn enthält ... ein Erkennendes und ein Erkanntes: sonst wäre es kein Bewußtseyn.*" His use of the words *'Selbstbewußtseyn'* (self-consciousness) and *'Bewußtseyn'* (consciousness) confirms my interpretation.

12  The will "is the being-in-itself of every thing in the world, and is the sole kernel of every phenomenon." (W1: 118)

13  Schopenhauer concedes that "Plants have at most an extremely feeble analogue of consciousness" (W2: 142)— that is, raw experiential states.

14  Schopenhauer's overall argumentation is quite consistent with this, so it is difficult to imagine that he means anything else.

15  "self-consciousness could not exist if there were not in it a

known opposed to the knower and different therefrom." (W2: 202)

16 Here Schopenhauer uses the word 'consciousness' in the broader, modern sense of *phenomenal* consciousness (Block 1995)—i.e. 'what-it-is-likeness'—without self-reflection or an implied knower-known pair. In his original German, Schopenhauer (1859) writes *"pflanzenartiges Bewußtsein,"* thus still using the same unambiguous word, *'Bewusstsein.'*

17 Under the assumption of locality.

18 At the time of this writing the transcript of this talk was available online at: http://web.stanford.edu/~alinde/Spir Quest.doc. See page 12.

19 Remember that Schopenhauer explicitly rejects solipsism, arguing that such a view "could be found only in a madhouse" (W1: 104).

20 Schopenhauer himself uses the same example in a closely related discussion (W1: 116) about the difference between voluntary actions ensuing on 'motive' and involuntary actions ensuing on 'stimulus.' In the context of the present discussion, motive entails meta-consciousness and stimulus doesn't.

21 "the will attains to self-consciousness only in the individual, and thus knows itself directly only as the individual" (W2: 510).

22 The wording here is nuanced, so let us look at the original German (Schopenhauer 1859) in its syntactical context: *"... und da was der Wille will immer das Leben ist, eben weil dasselbe nichts weiter, als die Darstellung jenes Wollens für die Vorstellung ist; so ist ... ".* The word *'Darstellung'* is really only a synonym for *'Vorstellung'*—in the sense that Schopenhauer uses the latter, i.e. as 'representation'—presumably to avoid word repetition. So the passage translates literally as "... and since what the will always wants is life, simply because the latter is nothing other than the representation of that

will for the representation, so is ...". In other words, what we call life, living organisms, is a perceptual representation of the realized desire of the will to have perceptual representations.

# Bibliography

American Psychiatric Association (2013). *Diagnostic and Statistical Manual of Mental Disorders* (5th ed.). Washington, DC: American Psychiatric Publishing.

Ananthaswamy, A. (2011). Quantum magic trick shows reality is what you make it. *New Scientist*, June 22. [Online]. Available from: https://www.newscientist.com/article/dn20600-quantum-magic-trick-shows-reality-is-what-you-make-it/ [Accessed 6 January 2019].

Aspect, A., Grangier, P. and Roger, G. (1981). Experimental tests of realistic local theories via Bell's theorem. *Physical Review Letters*, 47 (7): 460-463.

Aspect, A., Dalibard, J. and Roger, G. (1982). Experimental test of Bell's inequalities using time-varying analyzers. *Physical Review Letters*, 49 (25): 1804-1807.

Aspect, A., Grangier, P. and Roger, G. (1982). Experimental realization of Einstein- Podolsky-Rosen-Bohm gedankenexperiment: A new violation of Bell's inequalities. *Physical Review Letters*, 49 (2): 91-94.

Atmanspacher, H. (2014). 20th century variants of dual-aspect thinking. *Mind and Matter*, 12 (2): 245-288.

Barrett, D. (1994). Dreams in dissociative disorders. *Dreaming*, 4 (3): 165-175.

Bell, J. (1964). On the Einstein Podolsky Rosen paradox. *Physics*, 1 (3): 195-200.

Bernstein, D. A. (2010). *Essentials of Psychology*. Boston, MA: Cengage Learning.

Black, D. W. and Grant, J. E. (2014). *The Essential Companion to the Diagnostic and Statistical Manual of Mental Disorders* (5th ed.). Washington, DC: American Psychiatric Publishing.

Block, N. (1995). On a confusion about a function of consciousness. *Behavioral and Brain Sciences*, 18: 227-287.

Bohm, D. (1952a). A suggested interpretation of the quantum theory in terms of "hidden" variables. I. *Physical Review*, 85: 166-179.

Bohm, D. (1952b). A suggested interpretation of the quantum theory in terms of "hidden" variables. II. *Physical Review*, 85: 180-193.

Buks, E. *et al.* (1998). Dephasing in electron interference by a 'which-path' detector. *Nature*, 391: 871-874.

Cartwright, J. (2007). Quantum physics says goodbye to reality. *IOP Physics World*, April 20. [Online]. Available from: https:// physicsworld.com/a/quantum-physics-says-goodbye-to- reality/ [Accessed 6 January 2019].

Chalmers, D. (1996). *The Conscious Mind: In Search of a Fundamental Theory*. Oxford, UK: Oxford University Press.

Chalmers, D. (2003). Consciousness and its place in nature. In: Stich, S. & Warfield, T. (eds.). *The Blackwell Guide to Philosophy of Mind*. Malden, MA: Blackwell.

Chalmers, D. (2016). The combination problem for panpsychism. In: Brüntrup, G. & Jaskolla, L. (eds.). *Panpsychism*. Oxford, UK: Oxford University Press.

Chalmers, D. (2018). Idealism and the mind-body problem. In: Seager, W. (ed.). *The Routledge Handbook of Panpsychism*. London, UK: Routledge.

Cleeremans, A. (2011). The radical plasticity thesis: How the brain learns to be conscious. *Frontiers in Psychology*, 2, article 86.

Dijksterhuis, A. and Nordgren, L. F. (2006). A theory of unconscious thought. *Perspectives on Psychological Science*, 1 (2): 95-109.

Durant, W. (2006). *The Story of Philosophy: The Lives and Opinions of the Great Philosophers*. New York, NY: Pocket Books.

Eagleman, D. M. (2011). *Incognito: The Secret Lives of the Brain*. New York, NY: Canongate.

Einstein, A., Podolsky, B. and Rosen, N. (1935). Can quantum-mechanical description of physical reality be considered complete? *Physical Review*, 47: 777-780.

Emerging Technology from the arXiv (2019). A quantum experiment suggests there's no such thing as objective reality. *MIT Technology Review*, 12 March. [Online]. Available from: https://www.technologyreview.com/s/613092/a-quantum-experiment-suggests-theres-no-such-thing-as-objective-reality/ [Accessed 23 March 2019].

Gillespie, A. (2007). The social basis of self-reflection. In: Valsiner, J. and Rosa, A. (eds.). *The Cambridge Handbook of Sociocultural Psychology*. New York, NY: Cambridge University Press, pp. 678-691.

Gröblacher, S. *et al.* (2007). An experimental test of non-local realism. *Nature*, 446: 871-875.

Gu, M. *et al.* (2009). More really is different. *Physica D: Nonlinear Phenomena*, 238 (9-10): 835-839.

Hensen, B. *et al.* (2015). Loophole-free Bell inequality violation using electron spins separated by 1.3 kilometres. *Nature*, 526: 682-686.

Janaway, C. (2002). *Schopenhauer: A Very Short Introduction*. Oxford, UK: Oxford University Press.

Jung, C. G. (2001). *Modern Man in Search of a Soul*. New York, NY: Routledge.

Kastrup, B. (2017a). Making sense of the mental universe. *Philosophy and Cosmology*, 19: 33-49.

Kastrup, B. (2017b). The quest to solve problems that don't exist: Thought artifacts in contemporary ontology. *Studia Humana*, 6 (4): 45-51.

Kastrup, B., Stapp, H. P. and Kafatos, M. C. (2018). Coming to grips with the implications of quantum mechanics. *Scientific American*, May 29. [Online]. Available from: https://blogs.scientificamerican.com/observations/coming-to-grips-with-the-implications-of-quantum-mechanics/ [Accessed 6 January 2019].

Kastrup, B. (2018a). The universe in consciousness. *Journal of Consciousness Studies*, 25 (5-6): 125-155.

Kastrup, B. (2018b). Conflating abstraction with empirical observation: The false mind-matter dichotomy. *Constructivist Foundations*, 13 (3): 341-361.

Kierkegaard, S. (author) and Lowrie, W. (translator) (2013). *Fear and Trembling and The Sickness unto Death*. Princeton, NJ: Princeton University Press.

Kim, Y.-H. *et al.* (2000). A delayed 'choice' quantum eraser. *Physical Review Letters*, 84: 1-5.

Koch, C. (2014). Is Consciousness Universal? Panpsychism, the ancient doctrine that consciousness is universal, offers some lessons in how to think about subjective experience today. *Scientific American Mind*, January 1. [Online]. Available from: https://www.scientificamerican.com/article/is-consciousness-universal/ [Accessed 18 January 2019].

Lapkiewicz, R. *et al.* (2011). Experimental non-classicality of an indivisible quantum system. *Nature*, 474: 490-493.

Laughlin, R. B. and Pines, D. (2000). The Theory of Everything. *Proceedings of the National Academy of Sciences of the United States of America*, 97 (1): 28-31.

Leggett, A. J. (2003). Nonlocal hidden-variable theories and quantum mechanics: An incompatibility theorem. *Foundations of Physics*, 33 (10): 1469-1493.

Ma, X.-S. *et al.* (2013). Quantum erasure with causally disconnected choice. *Proceedings of the National Academy of Sciences of the Unites States of America*, 110 (4): 1221-1226.

Manning, A. G. *et al.* (2015). Wheeler's delayed-choice gedanken experiment with a single atom. *Nature Physics*, 11: 539-542.

Motl, L. (2009). Bohmists & segregation of primitive and contextual observables. *The Reference Frame*, January 23. [Online]. Available from: https://motls.blogspot.com/2009/01/bohmists-segregation-of-primitive-and.html [Accessed 6 January 2019].

Nagasawa, Y. and Wager, K. (2016). Panpsychism and priority cosmopsychism. In: Brüntrup, G. and Jaskolla, L. (eds.). *Panpsychism*. Oxford, UK: Oxford University Press.

Nagel, T. (1974). What is it like to be a bat? *The Philosophical Review*, 83 (4): 435-450.

Neumann, E. (2014). *The Origins and History of Consciousness*. Princeton, NJ: Princeton University Press.

Neumann, J. von (2018). *Mathematical Foundations of Quantum Mechanics: New Edition*. Princeton, NJ: Princeton University Press.

Nietzsche, F. (1919). *Die fröhliche Wissenschaft*. Prague, Czech Republic: e-artnow.

Nietzsche, F. (author) and Kaufmann, W. (translator) (1974). *The Gay Science*. New York, NY: Vintage Books.

Nietzsche, F. (author) and Nauckhoff, J. (translator) (2001). *The Gay Science*. Cambridge, UK: Cambridge University Press.

Nietzsche, F. (author) and Common, T. (translator) (2006). *The Gay Science*. Mineola, NY: Dover Publications, Inc.

Nixon, G. M. (2010). From panexperientialism to conscious experience: The continuum of experience. *Journal of Consciousness Exploration and Research*, 1 (3): 215-233.

Prideaux, S. (2018). *I Am Dynamite! A Life of Nietzsche*. New York, NY: Tim Duggan Books.

Proietti, M. *et al.* (2019). Experimental rejection of observer-independence in the quantum world. *arXiv:1902.05080 [quant-ph]*. [Online]. Available from: https://arxiv.org/abs/1902.05080 [Accessed 23 March 2019].

Romero, J. *et al.* (2010). Violation of Leggett inequalities in orbital angular momentum subspaces. *New Journal of Physics*, 12: 123007.

Rovelli, C. (1996). Relational Quantum Mechanics. *International Journal of Theoretical Physics*, 35 (8): 1637-1678.

Schaffer, J. (2010). Monism: The priority of the whole. *Philosophical Review*, 119 (1): 31-76.

Schlumpf, Y. *et al.* (2014). Dissociative part-dependent resting-state activity in Dissociative Identity Disorder: A controlled fMRI perfusion study. *PloS ONE*, 9, doi: 10.1371/journal. pone.0098795.

Schooler, J. (2002). Re-representing consciousness: dissociations between experience and meta-consciousness. *Trends in Cognitive Science*, 6 (8): 339-344.

Schopenhauer, A. (1859). *Die Welt als Wille und Vorstellung.* Leipzig, Germany: F. A. Brockhaus.

Schopenhauer, A. (author), Haldane, R. B. and Kemp, J. (translators) (1909). *The World as Will and Idea.* London, UK: Kegan Paul, Trench, Trübner & Co.

Schopenhauer, A. (author) and Payne, E. F. J. (translator) (1969). *The World as Will and Representation.* New York, NY: Dover Publications, Inc.

Shannon, C. E. (1948). A mathematical theory of communication. *Bell System Technical Journal*, 27: 379-423 & 623-656.

Stapp, H. P. (2001). Quantum theory and the role of mind in nature. *Foundations of Physics*, 31 (10): 1465-1499.

Strasburger, H. and Waldvogel, B. (2015). Sight and blindness in the same person: Gating in the visual system. *PsyCh Journal*, 4 (4): 178-185.

Strawson, G. *et al.* (2006). *Consciousness and Its Place in Nature: Does Physicalism Entail Panpsychism?* Exeter, UK: Imprint Academic.

Streater, R. F. (2007). *Lost Causes in and beyond Physics.* Berlin, Germany: Springer-Verlag.

Tanner, M. (2001). *Nietzsche: A Very Short Introduction.* Oxford, UK: Oxford University Press.

The BIG Bell Test Collaboration (2018). Challenging local realism with human choices. *Nature*, 557: 212-216.

Tittel, W. *et al.* (1998). Violation of Bell inequalities by photons more than 10 km apart. *Physical Review Letters*, 81 (17): 3563-3566.

Tsuchiya, N. *et al.* (2015). No-report paradigms: Extracting the true neural correlates of consciousness. *Trends in Cognitive Science*, 19 (12): 757-770.

Valsiner, J. (1998). *The Guided Mind*. Cambridge, MA: Harvard University Press.

Vandenbroucke, A. *et al.* (2014). Seeing without knowing: Neural signatures of perceptual inference in the absence of report. *Journal of Cognitive Neuroscience*, 26 (5): 955-969.

W1: see Schopenhauer & Payne (1969), volume 1.

W2: see Schopenhauer & Payne (1969), volume 2.

Weihs, G. *et al.* (1998). Violation of Bell's inequality under strict Einstein locality conditions. *Physical Review Letters*, 81 (23): 5039-5043.

Wicks, R. (2017). Arthur Schopenhauer. In: Zalta, E. N. (ed.). *The Stanford Encyclopedia of Philosophy (Summer 2017 Edition)*. [Online]. Available from: https://plato.stanford.edu/archives/sum2017/entries/schopenhauer/.

Windt, J. M. and Metzinger, T. (2007). The philosophy of dreaming and self-consciousness: what happens to the experiential subject during the dream state? In: Barrett, D. and McNamara, P. (eds.). *The New Science of Dreaming*. Westport, CT: Praeger, pp. 193-247.

Young, J. (2017). Arthur Schopenhauer: The first European Buddhist. *The Times Literary Supplement*, 24 August. [Online]. Available from: https://www.the-tls.co.uk/articles/public/arthur-schopenhauer-footnotes-to-plato/amp/ [Accessed 5 December 2018].

# ACADEMIC AND SPECIALIST

## Why Materialism Is Baloney

How true skeptics know there is no death and fathom answers to life, the universe, and everything
Bernardo Kastrup
A hard-nosed, logical, and skeptic non-materialist metaphysics, according to which the body is in mind, not mind in the body.
Paperback: 978-1-78279-362-5 ebook: 978-1-78279-361-8

## The Fall

Steve Taylor
*The Fall* discusses human achievement versus the issues of war, patriarchy and social inequality.
Paperback: 978-1-78535-804-3 ebook: 978-1-78535-805-0

## Brief Peeks Beyond

Critical essays on metaphysics, neuroscience, free will, skepticism and culture
Bernardo Kastrup
An incisive, original, compelling alternative to current mainstream cultural views and assumptions.
Paperback: 978-1-78535-018-4 ebook: 978-1-78535-019-1

## Framespotting
Changing how you look at things changes how
you see them
Laurence & Alison Matthews
A punchy, upbeat guide to framespotting. Spot deceptions and
hidden assumptions; swap growth for growing up. See and be free.
Paperback: 978-1-78279-689-3 ebook: 978-1-78279-822-4

## Is There an Afterlife?
David Fontana
Is there an Afterlife? If so what is it like? How do Western ideas
of the afterlife compare with Eastern? David Fontana presents
the historical and contemporary evidence for survival of physical
death.
Paperback: 978-1-90381-690-5

## Nothing Matters
a book about nothing
Ronald Green
Thinking about Nothing opens the world to everything by
illuminating new angles to old problems and stimulating new
ways of thinking.
Paperback: 978-1-84694-707-0 ebook: 978-1-78099-016-3

## Panpsychism
The Philosophy of the Sensuous Cosmos
Peter Ells
Are free will and mind chimeras? This book, anti-materialistic
but respecting science, answers: No! Mind is foundational to all
existence.
Paperback: 978-1-84694-505-2 ebook: 978-1-78099-018-7

## Punk Science
Inside the Mind of God
Manjir Samanta-Laughton
Many have experienced unexplainable phenomena; God, psychic abilities, extraordinary healing and angelic encounters. Can cutting-edge science actually explain phenomena previously thought of as 'paranormal'?
Paperback: 978-1-90504-793-2

## The Vagabond Spirit of Poetry
Edward Clarke
Spend time with the wisest poets of the modern age and of the past, and let Edward Clarke remind you of the importance of poetry in our industrialized world.
Paperback: 978-1-78279-370-0 ebook: 978-1-78279-369-4

Readers of ebooks can buy or view any of these bestsellers by clicking on the live link in the title. Most titles are published in paperback and as an ebook. Paperbacks are available in traditional bookshops. Both print and ebook formats are available online.
Find more titles and sign up to our readers' newsletter at
http://www.johnhuntpublishing.com/non-fiction
Follow us on Facebook at
https://www.facebook.com/JHPNonFiction
and Twitter at https://twitter.com/JHPNonFiction